EVOLUTION
and the
MOLECULAR REVOLUTION

EDITED BY

CHARLES R. MARSHALL

J. WILLIAM SCHOPF

JONES AND BARTLETT PUBLISHERS

Sudbury, Massachusetts

Boston   **London**   **Singapore**

Editorial, Sales, and Customer Service Offices
Jones and Bartlett Publishers
40 Tall Pine Drive
Sudbury, MA 01776
(508)443-5000
(800)832-0034

Jones and Bartlett Publishers International
7 Melrose Terrace
London W6 7RL
England

Library of Congress Cataloging-in-Publication Data

Evolution and the molecular revolution / edited by Charles R.
    Marshall, J. William Schopf.
        p. cm.
    Includes bibliographical references and index.
    ISBN 0-86720-910-0
    1. Evolution (Biology)   2. Molecular biology.   3. Genetics.
I. Marshall, Charles Richard, 1961–   . II. Schopf, J. William,
1941–   .
    QH366.2.E8495   1995
    575 — dc20   95–18551
                                                                                       CIP

Acquisitions Editors: Arthur Bartlett/David Phanco
Production Editor: Nadine Fitzwilliam
Manufacturing Buyer: Dana L. Cerrito
Typesetting: University Graphics, Inc.
Printing and Binding: Malloy Lithographing, Inc.
Cover Printing: Malloy Lithography, Inc.

A Contribution of the IGPP Center for the Study of Evolution and the Origin of Life (CSEOL), University of California, Los Angeles

Photos for the Biographical Sketches courtesy of Richard Mantonya.

Printed in the United States of America
99 98 97 96 95   10 9 8 7 6 5 4 3 2 1

CONTENTS

CHAPTER 3

GENES, SEQUENCES, AND CLOCKS: MOLECULAR CLUES TO THE HISTORY OF LIFE   53

Bruce Runnegar

CHAPTER 6

FROM MOLECULAR EVOLUTION TO BIOMEDICAL RESEARCH: THE CASE OF CHARLES DARWIN AND CHAGAS' DISEASE   125

Larry Simpson

PREFACE

Evolutionary biology is the bedrock of an enormous diversity of fields—evolution is the "GUT," the Grand Unifying Theory, that links together *all* life sciences. From the days of Darwin, insight into life's history has been afforded by studies of organismal anatomy, morphology, and embryology; of population biology, ecology, and biodiversity; of the fossil record, and its unique contribution to deciphering biotic development over geologic periods of time. But science, like the history of life, is far from static, and as the molecular revolution has swept the life sciences in recent years, the new data and knowledge of molecular biology have been added to the evolutionary story—evolutionary theory, itself, continues to evolve.

To gain a state-of-the-art introduction to the impact of this ongoing revolution, more than 2,000 students, faculty, and other members of the UCLA community gathered on March 18, 1994, to attend a day-long symposium called "Evolution and the Molecular Revolution," convened by the Center for the Study of Evolution and the Origin of Life at the University of California, Los Angeles. This book makes available the proceedings of that symposium.

Presented at a level appropriate for use in first- or second-year college coursework, the six chapters of this book give an overview of our current understanding of the evolutionary process, and show, through a selection of in-depth examples, some of the most exciting advances that molecular biology brings to the study of evolution. Coverage is not intended to be comprehensive; indeed, there is no way that it could be, given the pervasive influence of the molecular revolution on present-day evolutionary science. All topics covered focus on one overriding theme—the contribution of molecular biology to modern evolutionary thought—and all are designed to be accessible to those not yet fully familiar with the organization and intriguing intricacies of the fascinating, submicroscopic, biomolecular world.

Biographical sketches of the authors of the six chapters follow.

Charles R. Marshall
J. William Schopf
August, 1995

BIOGRAPHICAL
SKETCHES

Chapter 1: Darwinism in an Age of Molecular Revolution
CHARLES R. MARSHALL

Charles R. Marshall

Born in Canberra, Australia, Dr. Marshall received his undergraduate training at the Australian National University and his Ph.D. degree in 1989 from the Committee on Evolutionary Biology at the University of Chicago. After serving as a National Institutes of Health Post-Doctoral Fellow at Indiana University, Bloomington, he joined the faculty at the University of California, Los Angeles, where he is a member of the Department of Earth and Space Sciences, the Molecular Biology Institute, and the Institute of Geophysics and Planetary Physics. A recent recipient of a National Science Foundation Young Investigator Award, Professor Marshall is not only a superb scientist but a gifted teacher as well, admired by both his students and colleagues for his enthusiasm, clarity of expression, and depth of knowledge. Co-editor of this volume, Dr. Marshall is representative of a new generation of evolutionary biologists with primary training in mathematics, paleontology, evolutionary biology, and molecular biology. His research interests include fossil lungfish, sand dollars, mathematical models of the incompleteness of the fossil record, and molecular approaches in systematics and paleontology.

Chapter 2: How Did the Molecular Revolution Start? What Makes Evolution Happen?
THOMAS H. JUKES

Thomas H. Jukes

Born in England, Dr. Jukes received his Ph.D. degree in biochemistry, in 1933, from the University of Toronto, and has been awarded an Honorary D.Sc. degree by the University of Guelph, Ontario. He is Professor Emeritus in the Departments of Integrative Biology and of Nutritional Sciences, as well as a research chemist at the Space Sciences Laboratory, at the University of California, Berkeley. Professor Jukes has been elected to honorary life membership in the American Society of Animal Science and is recipient of the Borden Award in Poultry Nutrition, the Cain Memorial Award of the American Association for Cancer Research, and the Klaus Schwarz Commemorative Medal of the Association of Bioinorganic Scientists. He is the author of more than 500 scientific articles and three books, as well as editor of three other major volumes. Dr. Jukes has been a leader in the theoretical and practical applications of our understanding of the molecular basis of life. His areas of research include studies of the amino acid code, protein chemistry, and molecular evolution, as well as extensive investigations of diverse biomolecules, including the vitamin B complex, and of human nutrition.

Chapter 3: Genes, Sequences, and Clocks: Molecular Clues to the History of Life

BRUCE RUNNEGAR

Born and educated in Australia, Dr. Runnegar has received both a Ph.D. degree, in 1967, and a D.Sc. degree from the University of Queensland. In Australia, he has served as a member of the faculty at the University of Queensland and at the University of New England, where he was awarded a Personal Chair in Geology. He has also been a Visiting Professional Lecturer at George Washington University, and is an Honorary Research Associate at the U.S. National Museum of Natural History, Smithsonian Institution. He is currently a member of the Department of Earth and Space Sciences, the Molecular Biology Institute, and the Institute of Geophysics and Planetary Physics at the University of California, Los Angeles. A Fellow of the Australian Academy of Science, Dr. Runnegar's broad interests in paleontology have involved him in field and laboratory work in 21 countries. An authority on fossil molluscs, the Permian Period of geological time, and the oldest known multicellular fossils, he is also a leader in the emerging field of molecular paleontology, in which his interests are the use of nucleotide data to reconstruct the history of life, biomineralization, and studies of biomolecules such as collagen and hemoglobin.

Bruce Runnegar

Chapter 4: Metabolic Memories of Earth's Earliest Biosphere

J. WILLIAM SCHOPF

Dr. Schopf is a member of the Department of Earth and Space Sciences, the Molecular Biology Institute, and the Institute of Geophysics and Planetary Physics (IGPP) at the University of California, Los Angeles, as well as Director of the IGPP Center for the Study of Evolution and the Origin of Life (CSEOL). He received his Ph.D. degree from Harvard University in 1968. In addition to being co-editor of this volume, he has edited two previous publications, *Major Events in the History of Life* and *Creative Evolution?!*, resulting from CSEOL-sponsored public symposia, and two prize-winning monographs focusing on the earliest history of life on Earth. A member of the American Philosophical Society and of the American Academy of Arts and Sciences, Professor Schopf is recipient of two Guggenheim Fellowships as well as the Alan T. Waterman Award of the National Science Board, the Mary Clark Thompson Medal of the U.S. National Academy of Sciences, and the A. I. Oparin Medal of the International Society for the Study of the Origin of Life. Discoverer of the oldest fossils now known, he has carried out geological field studies in Africa, Asia, Australia, Europe, and North and South America.

J. William Schopf

Chapter 5: Homeotic Genes: Explaining the Evolution of Body Plans

E.M. DE ROBERTIS

Dr. De Robertis is the Norman Sprague Professor of Molecular Oncology and Professor of Biological Chemistry and Member of the Molecular Biology Institute, at the University of California, Los Angeles. He has also held positions at the Medical Research Council, Cambridge University, and at the University of Basel, Switzerland. He is an elected member of the European Molecular Biology Organization and has been a visiting Professor at academic and research institutions in Chile, Italy, and France. Dr. De Robertis received his Doctorate of Medicine at the University of Uruguay and his Ph.D. degree in biochemistry, in 1974, from the Faculty of Sciences, Buenos Aires, Argentina. Honored from the earliest days of his career, he has been awarded the Gold Medal as Outstanding Graduate in his medical school class at the University of Uruguay, and the Konex Foundation Prize, Buenos Aires, Argentina, for exceptional achievement in biomedical sciences during the decade 1983–1993. Among his numerous distinctions, Dr. De Robertis was the first to identify homeobox genes in vertebrates. His research interests are centered on understanding the genetic basis of the development of vertebrate animals.

E.M. De Robertis

Chapter 6: From Molecular Research to Biomedical Research: The Case of Charles Darwin and Chagas' Disease
LARRY SIMPSON

Larry Simpson

Dr. Simpson is a Professor in the Departments of Biology and of Medical Microbiology and Immunology, a member of the Molecular Biology Institute, of the Jonsson Cancer Center, and an Investigator at the Howard Hughes Institute of the University of California, Los Angeles. He received his Ph.D. in parasitology in 1967 from Rockefeller University. An outstanding experimentalist, he is recipient of the Seymour H. Hutner Award of the Society of Protozoologists and is an elected Fellow of the American Association for the Advancement of Science. He is also a dedicated and outstanding teacher, responsible for the UCLA Life Sciences Computing Facility, and has been honored as a Boyce-Thompson Distinguished Lecturer at Cornell University and a recipient of the Distinguished Faculty Teaching Award at UCLA. Dr. Simpson's research is centered on the mitochondrial genome of trypanosome parasites. He is not only a pacesetting researcher in the identification and characterization of novel genetic phenomena found in these kinetoplastid protozoans, but is also a leader in the use of these phenomena to diagnose human diseases caused by these widespread parasites.

CHAPTER 1

DARWINISM IN AN AGE OF MOLECULAR REVOLUTION

∎

Charles R. Marshall*

∎

INTRODUCTION

This chapter introduces Darwin's theory, or more correctly theories (Mayr, 1991), of evolution and discusses the types of contributions the molecular revolution has made to our understanding of evolution. In so doing, the chapter also provides an introduction to the rest of the book.

The core of Darwin's theories of evolution is remarkably simple. The first part of the chapter summarizes this core and some of its more important ramifications. The chapter also provides a brief review of how Darwin's theories are now viewed, 135 years after the publication of his book, *On the Origin of Species*.

The molecular revolution has been a genuine revolution, changing the face of almost every subdiscipline of biological science. It has been so pervasive that a comprehensive survey of the molecular revolution is not possible, nor desirable, in a book such as this. Instead, this chapter emphasizes the more recent aspects of the molecular revolution that have been among the most important in deepening our understanding of the evolution of life.

∎

DARWIN'S EVOLUTION

In 1859, Charles Darwin (Figure 1.1) published *On the Origin of Species*. Arguing against the idea that each species was created independently, Darwin outlined a compelling case for the origin of new species from previously existing species. Darwin's monograph is unusual for a revolutionary book, in that it is easily understood by those without specialized training, and because of the simplicity of its central tenets. In fact, the core of his argument is encapsulated in the book's Introduction [emphasis added]:

*Department of Earth and Space Sciences, Institute of Geophysics and Planetary Physics, and Molecular Biology Institute, University of California, Los Angeles, CA 90095.

FIGURE 1.1

Charles Darwin, around 1856, a few years before publication of *On the Origin of Species.*

As many more individuals of each species are born than can possibly survive; and as, consequently, there is a frequently recurring struggle for existence, it follows that any being, if it vary however slightly in any manner profitable to itself, under the complex and sometimes varying conditions of life, will have a better chance of surviving, and thus be **naturally selected**. From the strong principle of inheritance, any selected variety will tend to propagate its new and modified form. (Darwin, 1859)

Alfred Russel Wallace (1823–1913), a naturalist studying plants and animals in the Malay Archipelago at about the time Darwin was writing *On the Origin of Species*, independently arrived at the same conclusion. Note that the driving force of the evolutionary process is the struggle for survival, and that this struggle is driven by the fecundity of nature—it is this feature, the propensity of nature to overreproduce, that provides a basic condition necessary for the operation of Darwin's and Wallace's natural selection. Darwin felt that his theories could satisfactorily explain adaptation and its maintenance, as well as the origin of organic diversity.

Darwin's logical argument, or **syllogism**, can be recast in a number of different ways. Box 1.1 provides one version. The term **phenotype** refers to the observable characteristics of an organism, including its morphological, biochemical, and behavioral traits. The phenotype is the result both of genetic information (the **genotype**) and of the interaction of the organism with the environment.

The syllogism is straightforward. However, in Darwin's day it was not clear whether, or to what degree, the conditions of his logical argument were satisfied in nature. Much of *On the Origin of Species* is devoted to discussing the extent to which each of the conditions listed actually does exist in the natural world. There is perhaps one other condition that should be added to the syllogism: Local environments must vary over time to some degree. The term "environment" in this context has a very wide meaning, including not only physical conditions such as rainfall, temperature, seasonality, and so forth, but ecologic (for example, animal–animal, animal–plant) interactions as well. It should be noted that to a large degree organisms define and create their own environments; they are not just objects passively affected by environmental influences (Lewontin, 1983).

Darwin felt that what he termed natural selection was the most important cause of modification. Perhaps the most cited example of the operation of this natural selection is the explanation he offered for the origin of the long neck in the giraffe (Figure 1.2). Assuming that there is variation in neck length, and that this variation is, at least in part, heritable, Darwin argued that under the circumstances of low rainfall and drought, giraffes with slightly longer necks would have access to food not reached by others and, thus, the likelihood of survival of such long-necked individuals would be enhanced. It is important to note that the selective pressure for longer necks need not be applied continuously, or indeed even often—all that is required are occasional droughts to effect an increase in neck length over many generations. To establish that there is indeed a struggle for survival during drought, Darwin gave the example of the demise of the Niata cattle of South Amer-

BOX 1.1

Darwin's Syllogism.

Condition 1:	*If there is a*	Struggle for Existence, *and*
Condition 2:	*If the*	Phenotype Variable, *and*
Condition 3:	*If*	Existence Depends (at least in part) on the Phenotype, *and*
Condition 4:	*If*	Phenotypic Attributes are (at least in part) Heritable,
	Then	There will be Descent with Modification *that is,*
		Transformation of Species over Time

FIGURE 1.2

Giraffes in the savanna of East Africa. (Photograph provided by Kelly West, University of California, Los Angeles.)

ica during drought. Unlike horses and common cattle, Niata cattle cannot feed on twigs of trees, reeds, and so forth, and will starve during drought if not fed by their owners.

Darwin's interpretation of the evolution of the giraffe's neck also became a focus of criticism. For example, the English biologist George Jackson St. Mivart (1827–1900) argued that if a long neck were so good for a giraffe, why is such a neck not also beneficial to all other beasts that feed on vegetation? Darwin replied that selection for increased neck length operates only on the tallest animals; for example, a sheep with a slightly longer neck gains no advantage if its world is populated with cows and horses that are much taller than itself: "The competition of browsing on higher branches of the acacias and other trees must be between giraffe and giraffe, not with other ungulate animals" (Darwin, 1859, p. 193).

A more telling criticism highlights the lack of knowledge of the material basis of **heredity** at that time. In 1867, Fleeming Jenkin pointed out that any favorable mutation would soon become unimportant, swamped out by more normal organisms as the single mutant individual bred with other members of the **population.** This criticism would be valid if the hereditary material were infinitely divisible—for example, if it were like a droplet of wine diluted in a glass of water. Darwin certainly took Jenkin's criticism seriously because his model of inheritance was a blending one. However, with the rediscovery of Mendel and the development of population genetics by Wright, Fisher, and Haldane this issue was resolved. We now know that the hereditary material is *not* infinitely divisible or dilutable; instead, it is particulate in nature (made up of what are called **genes**). The introduction of a new trait into a population is more analogous to the addition of a red marble to a bag of blue marbles.

The Principle of Frustration

If natural selection purges relatively less fit characteristics from a population of organisms, how do we explain the maintenance of apparently deleterious traits in some populations, such as the maintenance of the gene for sickle cell anemia in some populations of humans? In humans suffering from sickle cell anemia, the **hemoglobin** molecules attach to each other when oxygen levels are low, causing the red blood cells to become contorted in a banana- or sickle-shaped fashion. These deformed **erythrocytes** may become trapped in the **capillaries**, causing pain and inflammation. They may also become permanently sickled, and these cells tend to have a shorter than average lifetime—patients thus afflicted will become anemic and they often die.

In each human cell there are two copies of each **chromosome** (one derived from the mother, the other from the father), and there are therefore two copies of each **gene** (that portion of the chromosome that contains the coded information to make a protein or RNA product). There is usually more than one version, or **allele**, of each gene. Sickle cell anemia is caused by a particular allele of hemoglobin called **hemoglobin S**. Individuals who have the hemoglobin S allele on both chromosomes (that is, who are **homozygous** for the hemoglobin S allele) often die as children (although in some environmentally benign locales, particularly those having advanced medical facilities, homozygous carriers may survive into adulthood). Yet despite its obviously deleterious effects, the allele is quite prevalent in some human populations, notably those in equatorial Africa.

So, if Darwin was right, why isn't the allele eliminated by natural selection? The reason is that although there is indeed strong natural selection *against* hemoglobin S homozygotes, individuals having only one copy of the hemoglobin S gene (referred to as **heterozygotes**) have an increased resistance to another potentially lethal disease, namely, malaria. Thus, in equatorial Africa, as well as in other places where malaria-carrying mosquitoes abound, having the hemoglobin S allele is both an advantage, if heterozygous, and a disadvantage, if homozygous (Allison, 1956). In contrast, in regions that lack malaria, the allele responsible for sickle cell anemia has been, as expected, purged from human populations (Figure 1.3).

The story of sickle cell anemia demonstrates a most important feature of the evolutionary process—the **Principle of Frustration.** All organisms live in, and help create, a complex environment, and they all have complex needs. Optimization to meet one need may not be optimal for another. For most organisms, most of the time, there are likely to be many selective forces operating simultaneously. The Principle of Frustration states that the best compromise solution to the inevitably conflicting demands of each of the organism's requirements is unlikely to be optimal for any single function. In fact, this principle may explain why there are so many different types of organisms. With the operation

FIGURE 1.3

Humans carrying the sickle cell anemia allele inhabit areas of malaria infestation in Africa. The larger map shows the distribution and frequency of occurrence of the sickle cell anemia allele, hemoglobin S, in the human population. From Sickle Cells and Evolution, Allison. Copyright © 1956 by Scientific American, Inc. All rights reserved. The smaller map shows that these same areas are infested with malaria. (Modified from Ridley, 1993. Reprinted by permission of Blackwell Scientific Publications, Inc.)

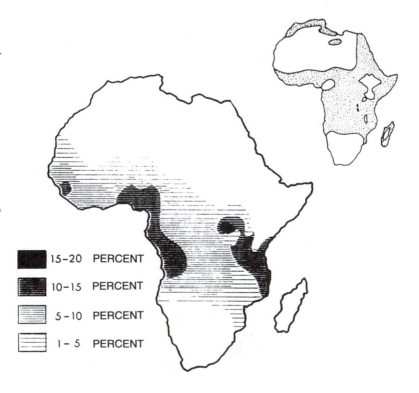

15-20 PERCENT

10-15 PERCENT

5-10 PERCENT

1-5 PERCENT

of conflicting selective forces, there are often many solutions that are more or less equally good in meeting the myriad of conflicting demands.

Niklas (1994), using a computer simulation of terrestrial plant morphology, has provided an elegant example of how the Principle of Frustration can lead to multiple successful solutions. Starting from a morphology similar to that of the earliest known land plants (Figure 1.4A), he considered the morphologies that would be most favored if there

A.   Ancestral Form

B.   Reproductive   Light Interception   Mechanical Stability
            Success

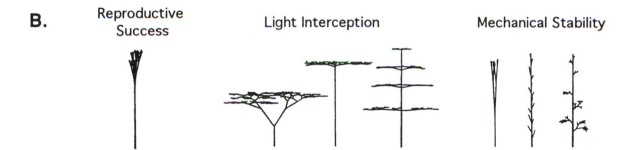

C.   Reproductive   +   Light   +   Mechanical
            Success   Interception   Stability

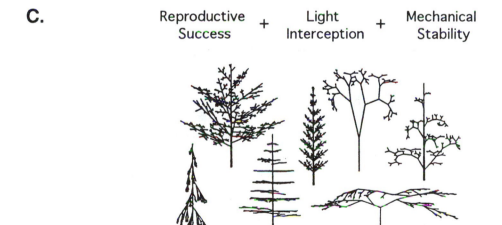

FIGURE 1.4

The evolutionary Principle of Frustration illustrated by Niklas's computer simulation of the evolution of land plants. *Part A*: Morphology used to begin the evolutionary simulations, based on the morphology of *Cooksonia*, the earliest known fossil vascular plant.

Part B: Optimal morphologies obtained when selection acted to optimize morphology to perform only a single function. *Part C*: Optimal forms obtained when selection acted to optimize morphological form to perform three functions simultaneously; a

greater range of equally optimal forms was generated, but each is suboptimal with respect to the forms that evolved to perform only a single function. (Modified from Niklas, 1994. Reprinted with permission.)

were selection for one of three different functions (Figure 1.4B). First, reproductive success was examined, in which the ability to produce many branches (and therefore many seeds) as high off the ground as possible (so that the seeds can disperse easily) was favored. Second, the ability to harvest light was examined, which favors the production of many nonoverlapping horizontal branches. And third, he examined the mechanical stability of these computer-generated plants, a character that, because of branch weight, favors morphologies having a limited number of branches, especially long horizontal ones. In each case, the resulting morphologies were certainly plantlike, but there were relatively few optimal solutions and none of the resulting trees looks especially like familiar trees such as maples or oaks (Figure 1.4B). Niklas then ran a simulation in which all three constraints were optimized simultaneously. This resulted in a dramatic increase in the range of optimal solutions, though none was optimal for any single function. Significantly, many of them look quite treelike (Figure 1.4C). It would appear, at least from this simulation, that the range of tree shapes with which we are familiar is the consequence of the simultaneous selection for several partially conflicting needs.

The Theory of Evolution Is Not like Physical Law

Darwin's contribution to our understanding of the world of nature is justifiably ranked alongside the works of Newton and Einstein (indeed, Darwin is buried alongside Newton in London's Westminster Abbey). However, as a cornerstone theory, the theory of evolution is different from the theoretical contributions of either Newton or Einstein. One of the remarkable features of the laws of physics is that the relationships encapsulated in equations, such as $e = mc^2$, are thought to hold virtually everywhere in the universe, both in space and time; gravitation and electromagnetic fields are properties of the universe that cannot easily be avoided. Given a set of initial conditions (e.g., a forming solar system of some specified mass, angular momentum, and so forth), and the appropriate laws (e.g., the Law of Conservation of Angular Momentum), the future development of that system can be predicted with some degree of certainty. In this sense, time is subordinate to physical law.

However, the theory of evolution does not state that evolution must occur; it is simply a list of conditions that, if met, will result in descent with modification. It does not preclude the occurrence of other processes that might be responsible for descent with modification. More important, the direction of evolution cannot be predicted in the same way that the future of (some) physical or chemical systems can be predicted. One of the most significant aspects of the evolutionary theory is that it implies that the development of life is **historically contingent**, that is, that what happens in the future depends very much on what is happening in the present and has happened in the past (Gould, 1989; Mayr, 1988).

For example, during the **Triassic Period** of geologic time, some 230 million years ago, the first turtles appeared. These are easily recognized because they possessed fully developed shells. However, no known Triassic turtle could retract its neck into its shell. By the **Cretaceous Period**, beginning about 100 million years later, two evolutionary lineages, each having a completely different neck-retraction mechanism, had become firmly established (Figure 1.5). In one of these, the neck was bent vertically (the so-called **cryptodires**, having a type of neck retraction that today is found only in turtles of the northern hemisphere); in the other, the neck was bent sideways (the **pleurodires**, found now only in the southern hemisphere).

With the appearance of a well-developed shell it is perhaps understandable (at least in retrospect) that neck retraction developed. First, because having a shell impeded mobility, it would be advantageous to be able to retract the head to escape pressure from predators. Second, the shell would offer a safe place into which the head could retract. How-

FIGURE 1.5

Evolution operates by opening and closing windows of opportunity. The extinction of turtles unable to retract their necks closed the window of opportunity that had led to the evolution of two different mechanisms for neck retraction. (Modified from Colbert and Morales, 1991.)

VERTICAL-NECK TURTLE

SIDE-NECK TURTLE

PRIMITIVE TURTLE - NO NECK RETRACTION

ever, once the ancestral stocks that were unable to retract their necks became extinct, developing a third neck-retraction mechanism would have been difficult. First, because neck-retraction mechanisms were already in place, there would not have been intense selection pressure to develop a new method of neck retraction, and second, it is difficult to envision how one of the known mechanisms of neck retraction could have been transformed into another. The evolution of the turtle shell opened opportunity for the evolution of neck-retraction mechanisms, and the disappearance of the ancestral forms that were unable to retract their necks evidently closed that evolutionary window.

Significance of Darwin

The idea that there has been descent with modification was not new with Darwin. This idea had already been gaining acceptance in the first half of the nineteenth century through the works of the great French naturalists such as Jean Baptiste Lamarck, Geoffroy Saint-Hilaire, and several others whose work Darwin reviews at the beginning of *On the Origin of Species*.

What was unique about Darwin's and Wallace's view of the process of evolution was that it was the first to provide a clear mechanism for the origin of species and their descent with modification. One of their most significant breakthroughs was their clear separation of the *source* of variation from the *needs* or requirements of the individual or-

BOX 1.2

Evolution Is a Two-Step Process.

Step 1:	Variation
Step 2:	Selection

ganisms. There is no divine or organismic "vote" on the variants produced; that is, variation is not directed by some outside agency, and the organism does not actively "try" to change or evolve. Mayr (1991) captured this key point in his characterization of evolution as a two-step process, as shown in Box 1.2.

The first five chapters of *On the Origin of Species* concern the existence and nature of variation, and the causes and character of natural selection. In discussions of variation, two significant points must be borne in mind (Box 1.3). It is quite a mouthful to say that "variation is random in the sense that the variants produced by nature do not anticipate the needs of the organism," so this is often abbreviated by stating simply that "variation is random." However, this abbreviation has been a source of misunderstanding. When read at face value, it would seem to suggest that any type of variant can be, and is, produced by nature. But this is not the case.

A delightful example of the limits on variation is Wayne's (1986) study of variation in domestic animals. There are a great many types of domestic dogs, ranging from great danes to pugs, but there is much less diversity among domestic cats. It could be argued that this difference has been brought about by humans, who have placed higher value on having many more types of dogs than cats, and this could well be the case. But there may be a much more profound reason for dogs showing more variation than cats. During its development, the relative proportions of a dog's skull change dramatically. Young dogs have broad and short skulls in comparison with those of adult dogs (Figure 1.6). In cats, however, the proportions of the face and head remain essentially constant during their growth to adulthood (Figure 1.6). This simple observation suggests that if there are large differences between the **morphologies** (the shape and form) of juveniles and adults in a species, a relatively greater range of varieties can be generated. This rule also holds for horses, the skulls of which show relatively little change of shape during growth and of which there are relatively few domestic varieties. On the other hand, pigs show dramatic face changes during their development and there are hundreds of different breeds. Wayne's analysis suggests that certain types of cat and horse skull morphologies are essentially impossible, not because they would be disadvantageous, but because the development of cats and horses proceeds in such a way that a whole suite of shapes simply cannot be readily generated. Clearly, the range of variation that can be produced by nature is not random with respect to the range of shapes and forms that can be imagined by humans!

The second significant point is that, in Darwin's day, the nature of heredity, let alone the source of heritable variation, was entirely unknown. In *On the Origin of Species*, Darwin's efforts were centered on demonstrating the existence of variation and describing the

BOX 1.3

Variation.

1. Random:	In the sense that the variants produced do not anticipate the "needs" of the organism	
2. Darwin:	"We are profoundly ignorant of the cause of each slight variation . . ." (Darwin, 1859, p. 171)	

FIGURE 1.6

Shape change during skull development of domestic dog and domestic cat. Greater shape change offers a larger range of variability for selection to occur in dogs compared with cats. This may be the reason for the larger number of breeds of dogs than of cats. (For ease of comparison, all skulls have been drawn at the same size. Modified from Wayne, 1986. Reprinted with permission.)

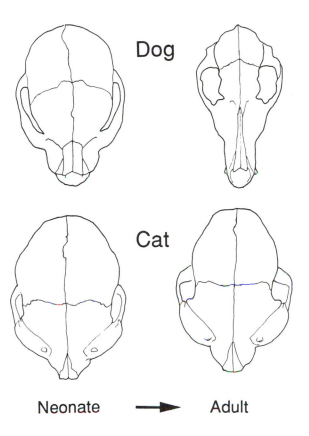

Dog

Cat

Neonate ➝ Adult

extents and types of variations that are present in organisms. Of course, the notion that there *are* variants is not surprising—after all, each of us looks different from everybody else. We still have much to learn about the nature of variation, and as was true in the nineteenth century, so during the twentieth century much more emphasis has been placed on understanding the selection stage of the two-step evolutionary process summarized in Box 1.2.

Perhaps the most controversial aspect of Darwin's theory was the idea that the direction of evolution is determined by natural selection—that it is not guided or directed by some transcendent force. Given the importance of this point, and Darwin's ignorance of the nature and cause of variation, he placed much more emphasis on the second step of the two-step process: Natural selection. His emphasis on natural selection had the unfortunate consequence of leading many to equate evolution with selection itself. However, these two should not be equated; as discussed below there can be selection without evolution, and evolution without selection. But first we need clear definitions of evolution and natural selection (Box 1.4):

The definition of Natural selection given in Box 1.4 requires some comment. Natural selection can be said to have operated if the survival of an organism depended, at least in

BOX 1.4

Definitions.

Evolution:	Descent with Modification
Natural Selection:	Preferential survival of individuals having profitable (advantageous) variations

part, on the presence of some profitable characteristic or trait not found in other members of the same species. It is important to recognize that, as used here, the word "profitable" is not meant to invoke notions of exploitation or selfishness. Rather, it is a value-free term in that it pertains to any characteristic that improves the chance of an organism producing offspring for the next generation. Profitable (advantageous) characteristics can in fact be selfish (the ability to wrestle away limited resources from others, for example), but they can also be innocuous (the ability to successfully ward off predators), or may even be cooperative (the ability to form groups in order to gain protection or to gather food).

Natural Selection without Evolution

Imagine a situation in which every generation had a few percent of all offspring born with some sort of deformation of their limbs. Presumably, these malformed individuals would be selected against, because they would be unable to move (to escape predators, for example) as easily as other members of the population. Here natural selection operates, but there is no evolution; that is, there is no descent with modification. In any situation in which the average phenotype (the normal characteristics of the species) is most suited to the environment, natural selection will favor the existing average phenotype at the expense of other unusual phenotypes that differ from the average. This type of selection is termed **stabilizing selection**. We encountered stabilizing selection earlier in this chapter. In the case of sickle cell anemia, discussed above, the intermediate phenotype (produced by the heterozygous genotype) was of selective advantage in malaria-infested regions; humans homozygous for the sickle cell anemia allele are selected against, as are those who lack the allele altogether. Selection without evolution is commonplace in the natural world.

Evolution without Selection

Figure 1.7 shows, in schematic form, a population tree, showing the relationship between parents in a population and their offspring. At the beginning of the sequence (shown at the bottom of the figure) there are 10 individuals, each able to reproduce asexually (that is, to divide and produce up to two offspring, each an exact copy of the parent). In each generation (represented by each row of circles), the organisms either become extinct (that is, they fail to reproduce); they reproduce themselves (one offspring survives); or they double in number (two offspring survive). After sixteen generations (shown at the top of the figure), there has been descent with modification in the population. Offspring have survived from only two of the original ten lineages, and members of the lineage represented by the black circles have increased in number to dominate the group. Moreover, there appears to have been an interesting "evolutionary dynamic"—early extinction of some lineages, followed by an especially successful evolutionary rebound by the black progeny. It would be tempting to conclude that the original black individual (at the bottom of the figure) had some sort of selective advantage over organisms of the other lineages—certainly, a pattern like that shown in Figure 1.7 could be the consequence of a selective process.

Surprisingly, however, the pattern shown in Figure 1.7 was generated by a purely random process. To make Figure 1.7, I first drew the ten circles of the first generation. I then pulled two pennies from my pocket and, to determine the fate of each individual in the next generation, I tossed the two coins. If two tails came up, the lineage became extinct. One head and one tail, a single offspring survived (the individual reproduced itself). Two heads, both offspring survived. I then repeated this procedure for each surviving individ-

FIGURE 1.7

Evolution without natural selection. The lineage represented by black circles is more successful than that of the white lineages or the stippled lineage. However, this success is not from selection for some intrinsically superior characteristic but, in this random simulation (see text), from chance alone.

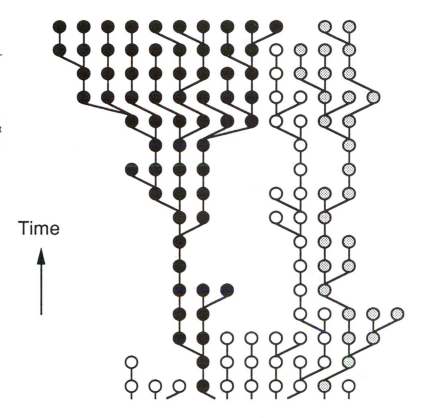

Time

ual in each successive generation. There was evolution (descent with modification), but because this was an entirely random process, there was no selection.

Does evolution without selection occur in nature? By the middle of the twentieth century, the answer to this question for the phenotype was no, or at least, only rarely. Evolution without selection was widely regarded as implausible. Reasoning that if this were true for the phenotype, it must also apply to the genotype, the Alexander Agassiz Professor of Vertebrate Paleontology at Harvard University, George Gaylord Simpson, declared that completely **neutral genes** or **alleles**—that is, genes or alleles that confer no selective advantage or disadvantage whatever—must be very rare, if they exist at all (Simpson, 1964). Most felt that the molecular makeup of an organism was likely to be so important to the survival of each individual that all portions of all molecules must be under selective constraint, including those molecules that make up genes. However, this could hardly have been further from the truth. Starting with Kimura (1968), and followed by a seminal paper by King and Jukes (1969), it has been established that neutral alleles predominate in genomes. In Chapter 2, Jukes explores the contribution that our understanding of the genetic basis of heredity has made to deciphering the importance of selection in genotypic evolution.

Provocatively, King and Jukes entitled their paper *Non-Darwinian Evolution*, and in their very first sentence they asserted that "Darwinism is so well established that it is difficult to think of evolution except in terms of selection for desirable characteristics and advantageous genes."

Here is reflected the prevailing spirit of the time, the view that Darwinian evolution and natural selection were one and the same, that one could not occur without the other. However, Darwin, in the last sentence of his introduction to *On the Origin of Species*, states [emphasis added]: "Furthermore, I am convinced that Natural Selection has been the most important, **but not the exclusive**, means of modification."

Thus, Darwin explicitly recognized that evolution can occur without selection (though he did not have neutral alleles in mind when writing this statement!). In fact, molecular evolution is *not* non-Darwinian, although evolution without selection is probably much more common at the genotypic than at the phenotypic level. Darwin's syllogism, his central logical argument, and its pervasive implications, have not been affected by the new insights provided by the molecular revolution (Brunk, 1991).

Explaining Today's Biosphere—Beyond the Darwinian Syllogism?

Among large **vertebrates** in today's terrestrial ecosystems, mammals are dominant. Why is this so? Why not an abundance of large reptiles, such as crocodiles, lizards, or even dinosaurs? Of course, if one were transported back to the **Mesozoic Era** (some 65 million to 250 million years ago) there would be no large mammals; dinosaurs were dominant at that time. Why the difference between then and now? A conventional explanation might be that mammals "out-competed" the dinosaurs, that mammals were somehow "better" than their reptilian relatives. This seems reasonable because the last dinosaurs are found in rocks 65 million years old (marking the end of the Cretaceous Period), and mammals first became abundant shortly after the dinosaur became extinct.

However, there is another possible explanation: Luck! It seems certain that Earth was blasted by a very large meteorite at essentially the same time as dinosaurs (and a great number of other groups) disappeared, at the end of the Cretaceous Period. The meteorite is thought to have been massive, having a diameter of approximately 10 km; its impact with the Earth would have ejected enough material into the atmosphere to cause drastic short-term changes in both the global climate and the chemistry of the world's oceans. The impact may have been the primary cause of the worldwide "**mass extinction**" that occurred at the end of the Cretaceous Period (Alvarez, 1983). Recently, the site of the impact was identified in Yucatan, Mexico—the huge Chicxulub crater, 180 km or more across (Sharpton et al., 1993).

It may be difficult to imagine that a meteorite impact could have truly global consequences. However, a back-of-the-envelope calculation shows that a large meteorite could have dramatic effects. As a standard for comparison, the Mt. St. Helens volcano ejected about 1 km^3 of material into the atmosphere when it exploded in 1980. Mt. Pinatubo, in the Philippines, was even more explosive, spewing out some 7 km^3 of rocky dust into the atmosphere when it erupted in 1991; it affected the world's climate for many months afterwards (for details, see the AGU Special Paper on Volcanism and Climate Change listed in the references for this chapter). It is estimated that the Cretaceous Period meteorite ejected as much as 22,000 km^3 of material into the atmosphere, certainly enough to affect the entire globe. The July 1994 bombardment of Jupiter by the fragments of the comet Shoemaker-Levy 9 should dispel any lingering doubts that collisions with large extraterrestrial objects can, and do, happen.

What are the ecological consequences of rapid climatic change? The fossil record, both at the end of the Cretaceous Period and at the end of the relatively recent last Ice Age (10,000 to 12,000 years ago), suggests that species with large body sizes are especially prone to becoming extinct during environmental crises. At the end of the Ice Age in North America, among other large animals, mammoths, giant ground sloths, camels, horses, lions, and saber-toothed cats all became extinct (possibly exterminated by humans). In fact, virtually all animals weighing more than about 50 kilograms (110 pounds) became extinct. Smaller beasts, such as coyotes, badgers, and raccoons faired much better and persist to the present. So also at the end of the Cretaceous Period all large species seem to have become extinct (Alvarez et al., 1980), whereas, for smaller species, some did and some did not.

Why large species seem to be more "extinction-prone" than small species at times of crisis is not known, but it is worth noting that at the end of the Cretaceous Period there were virtually no small dinosaurs, whereas all mammals at that time were small. Perhaps the main reason the dinosaurs became extinct is that there were simply no species with small body sizes present that might have survived. This idea, of course, is rather speculative. However, if it is correct, it has important implications for understanding the long-term history of biodiversity on Earth. In particular, it suggests that the reason why mammals are dominant now may be that the previously dominant vertebrates, the dinosaurs, were removed from the scene by the catastrophic impact, freeing up space that the mammals could then exploit. If so, mammals are not to be viewed as selectively superior to dinosaurs—they were simply lucky! Moreover, this hypothesis fits with the fact that mammals and dinosaurs coexisted for 160 million years before the end of the Cretaceous Period, with mammals remaining small and inconspicuous throughout this enormously long period.

Thus, while variation followed by selection (the operation of Darwin's syllogism) is most likely responsible for the appearance of new morphological traits, such as those features that make mammals different from dinosaurs, both the processes of Darwinian evolution and the occurrence of chance events appear to play a role in determining which groups survive over geologically long periods (tens to hundreds of millions of years) (Raup, 1991). Factors in addition to those included in Darwin's syllogism (Box 1.1) may also be important in determining the range of biodiversity of the modern world, or of that at any other time in Earth's history.

The effects on the **biota** of long-term geologic change should not be underestimated. For example, for the first 99% of their history, the small North American predators of the present day, such as the coyote and bobcat, coexisted with much larger predators, such as giant lions and saber-toothed cats. It is only in the last 10,000 to 12,000 years that these smaller predators have had North America to themselves, unhampered by competition with larger carnivores. Thus, many of the behavioral and morphological characteristics of these small currently living predators may be in part the product of nearly 2 million years of earlier competition with now extinct megacarnivores, a factor that must be taken into account in deciphering their evolutionary history. Without an appreciation of how radically, and how recently (in geologic time), the ecology of North America has changed, serious errors can be made in identifying the selective pressures that have shaped the characteristics of animals such as coyotes and bobcats (Van Valkenburgh and Hertel, 1993).

THE MOLECULAR REVOLUTION: RECOGNIZING SPECIES

We now turn to some of the major contributions molecular studies have made to our understanding of the history of life. Darwin discusses at some length the difficulties in distinguishing between populations that belong to different species and those that are simply different varieties of a single species. This difficulty is not surprising; gradations between varieties and species in a world with descent with modification are likely to be seen. In Darwin's time, as today, there is no universally agreed upon, comprehensive definition of a species. Nevertheless, for sexually reproducing organisms, one of the most widely accepted definitions is the **Biological Species Concept** proposed by Mayr (1942):

> Species are groups of actually or potentially interbreeding natural populations, which are reproductively isolated from other such groups.

Although applicable to many forms of life, this definition is not all-inclusive. For example, it obviously cannot be applied to organisms that reproduce asexually (most microbes and many protozoans and single-celled algae, for example), nor can it be applied directly to fossil species. In addition, and although it is a definition that works in theory, it is often not possible to know whether members of geographically isolated populations can actually mate. Other definitions of species have been proposed (summarized by Coyne, 1994), but none is any freer from difficulties.

Finding a comprehensive definition of species is complicated by other problems as well. For example, seven **subspecies** of the salamander *Ensatina eschscholtzii* (found in Washington, Oregon, and California) are recognized primarily by their coloration (Figure 1.8). In California, the salamanders primarily inhabit the foothills that surround the Central Valley (Figure 1.9). One subspecies (*Ensatina eschscholtzii eschscholtzii*; [*eschscholtzii* for short]; an unblotched form), is thought to have migrated down the coast, and is found in coastal southern California; another subspecies (*klauberi*, a blotched form) is thought to have migrated from the Sierra Nevada to the mountains of southern California. In most cases, adjacent subspecies show evidence of interbreeding: The blotched subspecies in mountainous southern California (*klauberi*) can interbreed with the subspecies occurring in the southern Sierran foothills to the north (*croceator*), and *croceator* can interbreed with the more northern California subspecies (*platensis*); *platensis* can interbreed with the central Californian coastal subspecies (*oregonensis* and *xanthoptica*), and these subspecies, in turn, can interbreed with the unblotched subspecies of coastal southern California (*eschscholtzii*). Because all these subspecies can interbreed, by Mayr's definition, they all fit into a single biological species.

The blotched (*klauberi*) and unblotched (*eschscholtzii*) subspecies that occupy southern California are in contact (Figure 1.9). Surprisingly, however, *klauberi* and *eschscholtzii* are unable to interbreed. By Mayr's definition, they should be assigned to different species. If one starts with the southwestern-most subspecies (*klauberi*) and goes counterclockwise around the Central Valley there appears to be one species of *Ensatina eschscholtzii*, but if one proceeds around the ring in the opposite direction, there appear to be two species. *Ensatina eschscholtzii* is a famous example of a so-called **ring species**, and it highlights some of the practical difficulties in recognizing and defining species. However, because there are sometimes fuzzy lines between species and varieties (subspecies), does not mean that the concept of a species should be abandoned—there are no sharp boundaries between the colors in a rainbow, but color is nevertheless an important tool when negotiating traffic signals! Weiner (1994) provides an excellent review of the evidence that natural selection and speciation are real processes.

FIGURE 1.8

Drawings of salamanders belonging to the ring species *Ensatina eschscholtzii*. The blotched form (right) occurs only east of the Californian Central Valley; the unblotched form (left) occurs west of the Central Valley. (Reprinted with permission from Moritz et al., 1992.)

FIGURE 1.9

Geographic distribution of the several subspecies belonging to the salamander ring species *Ensatina eschscholtzii*. Adjacent subspecies can interbreed, except for the two southern subspecies (*eschscholtzii* and *klauberi*) that are in contact in southern California. (Modified from Frolich, 1991. Reprinted by permission of Academic Press, Ltd.)

Ring species are relatively uncommon, but the same basic problem of distinguishing between varieties, subspecies, and species often plagues species recognition. Morphologically, it is difficult to distinguish between varieties and species. Molecules have provided an additional way of addressing the question of when a species is truly a species.

Mitochondrial DNA and Intraspecific Relationships

Virtually all cells of plants and animals contain energy-producing organelles known as **mitochondria**. Each mitochondrion contains its own circular piece of gene-containing DNA (**mtDNA**). Mitochondria are usually inherited from the mother. Over time, **mutations** (changes in the genes) occur, and all the descendants of the mother that first received the mutation inherit the mutant gene. (As discussed above, most of these changes are thought to be neutral, giving no advantage or disadvantage in the struggle for survival.) Thus, mutated mitochondria are like family names (except that they are inherited from the mother, rather than the father) and can be used to trace familial genealogical relationships. During the course of evolution, populations that are not in close contact tend to develop different mtDNA sequences. This property can be used to help distinguish populations of true, noninterbreeding species from those of interbreeding subspecies and varieties. In the case discussed earlier of the amphibian ring species, *Ensatina eschscholtzii*, mitochondrial DNA studies have confirmed that it is indeed a ring species (Moritz et al., 1992).

Conservation Biology and the Case of the Dusky Seaside Sparrow

The dusky seaside sparrow was first described in 1872 and was largely confined to Brevard County on the central Atlantic coast of Florida (the population labeled *nigrescens* in Figure 1.10A). In the early 1960s, flooding for mosquito control and conversion of wood-

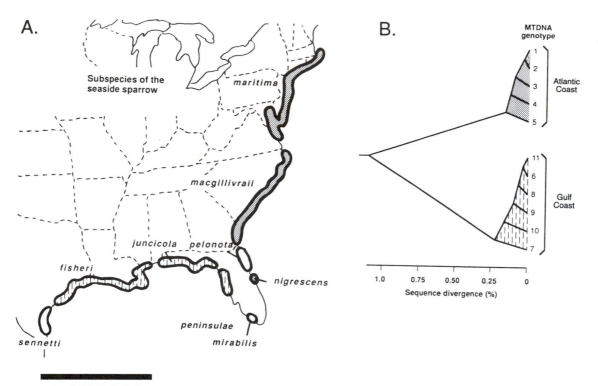

FIGURE 1.10

Geographic distribution and evolutionary tree of seaside sparrow subspecies. *Part A*: Subspecies populations range from the coast of the western Gulf of Mexico to the coast of the North Atlantic Ocean. The extinct dusky seaside sparrow is the population in Florida designated *nigrescens*. *Part B*: Evolutionary tree showing the relative degree of relatedness of seaside sparrow subspecies; numbers refer to individual mitochondrial types ("family names"). The greater the sequence divergence, the greater the genetic difference between populations. Although all Atlantic Coast populations are closely related, as are all Gulf Coast populations, these two sets of populations have diverged greatly. The extinct dusky seaside sparrow was genetically almost indistinguishable from the other Atlantic Coast seaside sparrows, but differed markedly from the Gulf Coast sparrows, including those of western Florida. (Modified from Avise and Nelson, 1989. Reprinted with permission. Copyright 1989 American Association for the Advancement of Science.)

lands into pasture caused a significant drop in population numbers of this sparrow and, by 1966, it was listed as an endangered species. By 1980, only six birds, all male, remained. Five of these were crossed with females obtained from a population of closely related sparrows inhabiting the western coast of Florida on the Gulf of Mexico. The last purebred dusky seaside sparrow died on June 16, 1987. Considerable effort had been spent trying to save the species, or to preserve at least some of its genetic diversity, by crossing the last survivors with the Gulf Coast sparrows. However, genetic analysis revealed that this effort may have been misspent.

When the mitochondria of the dusky seaside sparrow were compared with other sparrows from both the Gulf and Atlantic Coasts, two startling patterns were discovered (Avise and Nelson, 1989). First, dusky seaside sparrow mitochondria were remarkably similar to those of all other Atlantic Coast species. Second, their genetic makeup differed greatly from the mitochondria in populations from the Gulf of Mexico, including those in sparrows from western Florida (Figure 1.10B). Thus, it appears that the dusky seaside sparrow was not different from the other Atlantic species at all. Moreover, if its genetic diversity were to be preserved, the few remaining male birds should have been crossed with females from other conspecific Atlantic coast populations, not with those from populations from Florida's Gulf Coast. Loss of genetic diversity is certainly an important problem, but at a time when many bona fide species are threatened with extinction, mainly because of the activities of humans, it is crucial that the limited conservation resources available be concentrated on the most pressing cases. Molecular studies are playing an increasingly major role in conservation biology.

How Do Sea Turtles Find Their Nesting Grounds?

For most of their lives sea turtles live thousands of miles from their nesting grounds. An intriguing problem has been how they determine where to nest. Do females return to nest on the beach of their birth (a behavior known as **natal homing**), or do they follow experienced females to some other beach that the experienced females happen to select (**social facilitation**)? It has been impossible to trace the life of individual turtles from birth until they begin nesting because infant mortality is high, it takes many years for turtles to mature, and it has proven difficult to devise a tag that can be attached to juveniles that will survive until their adulthood.

However, this problem can be addressed by tracing female lineages through their mitochondrial DNAs. If natal homing is the primary mechanism of nesting site selection, members of the same lineage should return to the same site through the generations; there should be distinct types of mtDNA at each site, and they should differ from one beach to the next. If social facilitation is the primary mechanism of nest choice, various mtDNA types should be found across the range of nesting grounds, because females hatched on one beach may deposit eggs on many others.

Avise and his colleagues (Bowen et al., 1993) examined the mtDNA of loggerhead turtles in the northwestern Atlantic Ocean and the Mediterranean Sea. Of the five mtDNA "family names" found, there were two nesting localities in which only one of the five occurred; thus, evidently, nesting turtles return to the beach of their birth (Figure 1.11). However, there is some mixing of mtDNA types between beaches, so there is also apparently some degree of social facilitation, but this is relatively rare. These data have important conservation implications: If a nesting site is destroyed, females that hatched on that beach will not easily find their way to another beach to deposit eggs and that turtle population will most likely become extinct. Here, molecular data have not only answered an interesting biological question, but they have also suggested specific strategies that might be employed if the survival of the loggerhead turtle is valued.

THE MOLECULAR REVOLUTION: DETERMINING EVOLUTIONARY RELATIONSHIPS

Descent with modification (and the assumption that all living things share a common ancestor) implies that all species are related to each other. Hence, it should be possible to determine the genealogical relationships among species, that is, to reconstruct the Tree of Life. Traditionally, the differing degrees of morphological similarity among species were used to determine relative degrees of relatedness. **Systematists** (scientists concerned with the classification of species) search for shared evolutionary innovations (**synapomorphies**) as indicators of relationship. For example, in the late 1700s, the French comparative anatomist, Georges Cuvier (1769–1832), recognized that a very large and highly unusual fossil skull found in the Maastricht region (Figure 1.12A) of what is now the southernmost region of the Netherlands had affinities with lizards. The fossil belonged to a Cre-

FIGURE 1.11

Mitochondrial types (mtDNA "family names") of female loggerhead turtles surveyed at their nesting grounds. Each of five mtDNA types is designated by a letter (A, B, C, D, or E). If social facilitation were the primary mode of nest site selection, a single mitochondrial type would not be dominant on any individual beach, as is the case in southern Greece and in South Carolina and Georgia; the data support the hypothesis of natal homing (see text). (From Bowen et al., 1993. Reprinted by permission of Blackwell Scientific Publications, Inc.)

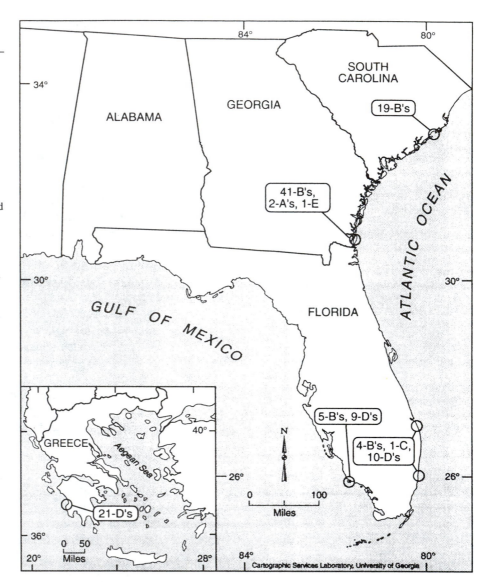

taceous marine reptile known as a **mosasaur**, and was identified as being related to modern **varanid** lizards, a group that includes the largest of all living lizards, the Komodo Dragon of Indonesia, as well as the goannas, or monitors, of Australia and southeast Asia. One of the major criteria for Cuvier's identification was a joint in the mosasaur's lower jaw, known to be present only in living varanids and fossil mosasaurs (Figure 1.12B). Thus, this jaw joint is a synapomorphy, an evolutionary innovation shared by these two types of reptiles.

The use of morphological data to infer evolutionary relationships has generally been very successful with vertebrate organisms, in part because of the complexity of their skeletons and soft tissues, and in part also, perhaps because humans as vertebrates have an uncommon interest in vertebrate anatomy. However, for animals without backbones, the **invertebrates**, and even more so for minute single-celled organisms such as bacteria, there is less morphology to work with and evolutionary relationships have been less well understood. With these groups the use of molecules has been revolutionary for our understanding of relationships.

Just as with morphological data, so with molecular data, unique evolutionary innovations (molecular synapomorphies) can be sought as indicators of relatedness (Box 1.5).

FIGURE 1.12

Determining the relationship of extinct mosasaurs to living varanid lizards. *Part A*: Depiction of the discovery of the skull of a giant seagoing Cretaceous mosasaur in a limestone quarry near Maastricht, Holland, in 1770. *Part B*: Skull of a varanid lizard showing the additional lower jaw joint present only in varanids and mosasaurs. (Modified from Benton, 1990. Courtesy of Peabody Museum of Natural History, Yale University.) This shared characteristic, a synapomorphy, indicates that mosasaurs and varanids such as the Komodo Dragon are more closely related to each other than to any other group of organisms.

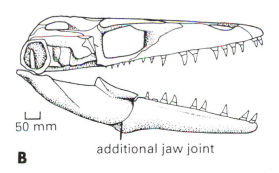

50 mm

additional jaw joint

B

An advantage of using molecular sequence data for reconstructing evolutionary trees is that, potentially, there are millions of synapomorphies waiting to be discovered. One of the most important molecules studied for this purpose is a ubiquitous and highly conserved molecule—that is, a molecule that has changed little during evolution—known as the **small subunit ribosomal RNA (rRNA)** (Figure 1.13). Figure 1.14 shows an **evolutionary tree** derived from the nucleotide sequences found in this molecule in a large number of different types of organisms. Several conclusions can be drawn from this tree: (1) Because rRNA occurs in all organisms except viruses, the degree of relatedness among every form of life can be determined by studies of this single type of molecule. Moreover, the universality of rRNA provides strong evidence that all life that exists today shared a common ancestor (of course, even stronger evidence comes from the fact that the genetic code is the same for all organisms). (2) the molecular differences within the animal and plant branches of the tree (stippled) are trivial in comparison with the molecular diversity exhibited by all life. (3) not only is there a great diversity within the bacteria, but they fall into one of two widely divergent groups, the **Eubacteria**, and the **Archaebacteria**.

There is also another conclusion that can be drawn from the Tree of Life shown in Figure 1.14. Most animal cells contain two discrete sources of DNA: Nuclear DNA, organized in the familiar chromosomes that make up most (>99.99%) of the cell's DNA, and mitochondria that contain their own DNA. Plant cells, in addition to nuclei and mitochondria, have a third source of DNA, **chloroplasts**, that house the machinery for **photosynthesis**. It has long been proposed that the ancestors of chloroplasts and mitochondria were once free-living bacteria, and that an early eukaryotic cell acquired these by engulfing them from their environment. The evolutionary tree in Figure 1.14 provides strong support for this hypothesis: The small subunit rDNA sequences of the corn mitochondria and chloroplasts (indicated by *) both are deeply embedded in the eubacterial

BOX 1.5

Use of DNA Sequences to Determine Evolutionary Relationships.

The DNA nucleotide sequence (A = adenine, T = thymine, C = cytosine, and G = guanine) for two small regions of the small subunit rRNA gene (Figure 1.13) for two mammals, two birds, and a **lobe-finned fish**†

Site Numbers:	920			980	
Mammal (Human)	CCGCC	..	TTTCG	GAACT	GAGGC
Mammal (Mouse)	CCGCC	..	TTTCG	GAACT	GAGGC
Bird (Chicken)	CCGCC	..	TTTCG	GAAAC	GGGGC
Bird (American Robin)	CCGCC	..	TTTCG	GAAAC	GGGGC
Lobe-finned fish (Coelacanth)	TCGCT	..	TTTCG	GAACT	GGGGC

If the lobe-finned fish (Coelacanth) sequence is assumed the primitive sequence, the sequences support this evolutionary tree:

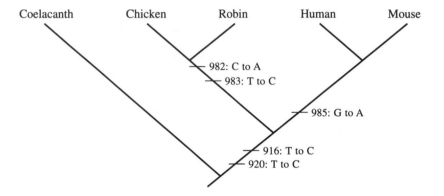

Each internal branch of the tree is supported by one or two unique evolutionary innovations (synapomorphies); for example, the human and mouse sequences both have A (adenine) at site 985, in contrast to all other species; this is a molecular synapomorphy uniting human and mouse to the exclusion of the other three species. The sequences in this example illustrate one of the ways DNA sequences can be used to reconstruct evolutionary relationships. The regions of the molecule were deliberately chosen to give consistent results; in reality, the analysis of DNA sequences is not so simple because there is usually conflicting information in any single data set. For example, at site 1068 (not shown here), the robin, human, and mouse share a G (guanine), while the chicken and coelacanth share an A. Based only on this site the robin would appear to be more closely related to the two mammals than it is to a chicken! One of the greatest challenges in the use of molecules to reconstruct evolutionary histories is to develop robust methods by which to deal with the conflicting information contained in molecular data sets.

†Hedges et al., 1990.

branch of the Tree of Life, not far from the well-studied bacterium, *Escherichia coli*, but very distant from the corn nuclear DNA (indicated by #).

Alternative Approaches to Molecular Phylogenetics

Although nucleotide sequences of RNA and DNA are most commonly used to reconstruct evolutionary relationships, there is other information in the genome that can be used to

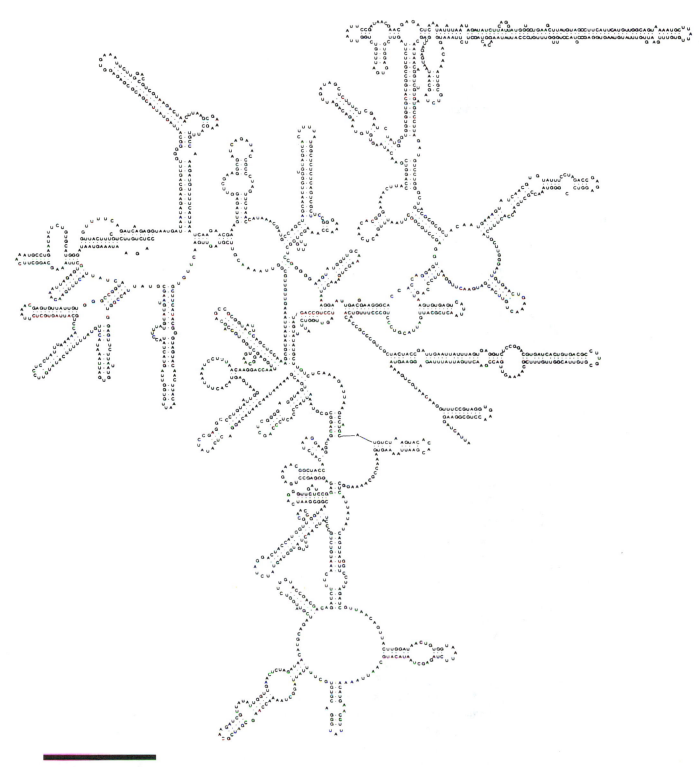

FIGURE 1.13

Schematic representation of a ribosomal RNA (rRNA) molecule. The 1995 nucleotides of the small subunit ribosomal RNA (SSU rRNA) from the fruitfly *Drosophila* *melanogaster* are arranged to show the secondary structure (the pairing of specific regions) of the molecule. This secondary structure is very similar in all organisms, from bacteria to humans. (Modified from De Rijk et al., 1992. Reprinted by permission of Oxford University Press.)

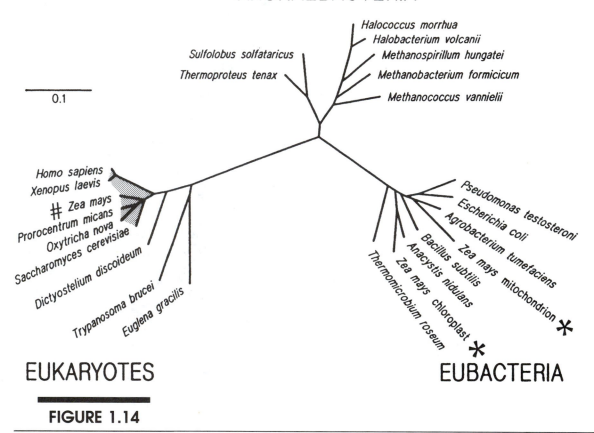

ARCHAEBACTERIA

Halococcus morrhua
Halobacterium volcanii
Methanospirillum hungatei
Methanobacterium formicicum
Methanococcus vannielii

Sulfolobus solfataricus
Thermoproteus tenax

0.1

Homo sapiens
Xenopus laevis
Zea mays
Prorocentrum micans
Oxytricha nova
Saccharomyces cerevisiae
Dictyostelium discoideum
Trypanosoma brucei
Euglena gracilis

Pseudomonas testosteroni
Escherichia coli
Agrobacterium tumefaciens
Zea mays mitochondrion
Bacillus subtilis
Anacystis nidulans
Zea mays chloroplast
Thermomicrobium roseum

EUKARYOTES

EUBACTERIA

FIGURE 1.14

The universal Tree of Life. The relationships shown were derived from the analysis of nucleotide sequences of the small subunit ribosomal RNA molecule (see Figure 1.13) extracted from diverse types of organisms. Animals (represented by *Homo sapiens* and the frog *Xenopus laevis*), plants (corn *Zea mays*), and fungi (the yeast *Saccharomyces cerevisiae*) occupy just a small corner of the tree (stippled). Note that the mitochondrial and chloroplast sequences from corn (*) are not closely related to the sequence from the cell nucleus of corn (#), but are closely related to the Eubacteria. This provides powerful evidence that mitochondria and chloroplasts are evolutionary derivatives of bacteria that were engulfed by an ancestral eukaryotic cell. (Details of the relationships among the Archaebacteria may be incorrect (see Chapter 3). The root of this Tree of Life (not shown) probably lies in the region between the Eubacteria and Archaebacteria. Scale bar shows distances on tree that correspond to a 10% difference in nucleotide sequence. (Modified from Olsen, 1989. Reprinted with permission.)

infer **phylogenies** (evolutionary trees). In particular, the genome contains a record of rare genomic events that may have phylogenetic value.

Most genetic information is passed from one generation to the next through sexual or asexual reproduction. This standard form of inheritance is termed **vertical descent**. But, very occasionally, a piece of DNA is transmitted *between* different species. This type of rare genomic event is termed **horizontal transfer**. An example of horizontal transfer is the case of a 10-kilobase (kb) piece of DNA that belonged to a particular type of virus, called a **retrovirus**, that was transmitted from an ancestor of the baboon to an ancestor of many types of cats, including the black-footed cat, African wildcat, sand cat, domestic cat, European wildcat, and jungle cat (Benveniste, 1985) (Figure 1.15). All these cats have multiple copies of this retroviral DNA in their cells. However, no other member of the cat family (**felids**) has the sequence—it is not present in the leopard cat, bobcat, lion, caracal, and many others. Just as the joint in the lower jaw of the mosasaur indicated its

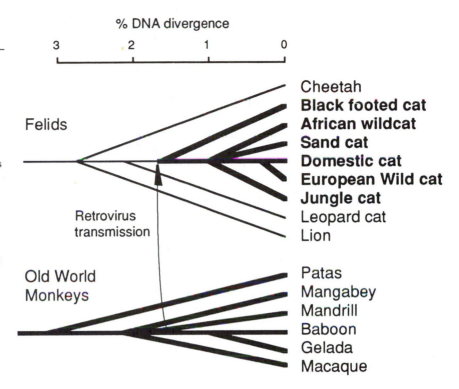

FIGURE 1.15

A 10-kb retroviral sequence transferred from a primate to an ancestral cat. This horizontal transfer is an excellent evolutionary marker indicating that six cats (names in bold type) are closely related. Species containing the retroviral sequence in their genomes are indicated by the thick lines. Relationships among the felids and Old World monkeys were inferred from measured differences in all sequences of DNA that occur once, or only a few times, in the genome using DNA-DNA hybridization (see Chapter 2). Branch lengths are approximately proportional to the amount of geologic time that has elapsed since the indicated divergences. (Modified from Benveniste, 1985. Reprinted with permission.)

evolutionary relationships with the varanids, the presence of this retrovirus sequence indicates that the six species of cat that have copies of the 10-kb piece of DNA are more closely related to each other than to any other felid. The shared presence of the primate retroviral sequence in their genomes is an excellent molecular synapomorphy for this group of cats.

Apparently, about 5 to 10 million years ago in North Africa, an ancestor of the cats that now share the retroviral sequence somehow became infected with the retrovirus from a primate that was ancestral to both the baboon and the gelada (Figure 1.15). Even though all primates, including humans, have copies of the retrovirus in their genomes, the cats seem to have received it from an ancestor of the baboon and gelada because the cat sequences show the greatest similarity to the retroviral sequences present in these two primates. It is possible that the infection occurred after the cat devoured the primate. This piece of DNA from the primate managed to incorporate itself into the DNA of the sperm or egg cells of the cat, and from there it was inherited vertically by all individuals descended from that initial cat. Although such horizontal transfers are rare, other examples are also known (Benveniste, 1985). However, the use of rare genomic events like this to reconstruct evolutionary relationships is limited because they are rare, and therefore they cannot be searched for systematically. In Chapter 3, Runnegar explores the use of molecular data to unravel the early history of life, and in Chapter 4, Schopf discusses the origin and evolution of the fundamental biochemical pathways used by all life.

THE MOLECULAR REVOLUTION: EXPLAINING MORPHOLOGICAL TRANSITIONS

A major area of current research is focused on the use of molecular data to reconstruct the Tree of Life, but what more can be learned once such a tree has been defined? One of the most interesting questions is how the various morphologies seen in extinct and extant organisms originated. As Gould (1980) pointed out in his essay on the panda's thumb, new structures generally do not originate *de novo* in evolution; they are usually modifications of previous structures.

One might be tempted to ask how did a chimpanzee become human? Or what were the steps taken to transform an arthropod into a dinosaur? These, however, are not good questions. Chimpanzees were not transformed into humans, nor arthropods into dinosaurs (Figure 1.16A). The species are related, certainly, but one did not evolve into the other. The real questions are what was the *most recent common ancestor* of the organisms of interest, and how did each evolving lineage acquire its unique characteristics since the time of that shared ancestor (Figure 1.16B)?

In some cases, comparative morphology provides a fairly complete picture of the nature of morphological innovation. For instance, the transformation of three of the bones of the lower jaw of the ancestors of living reptiles and mammals into the bones of the middle ear of extant mammals is relatively well described on the basis of the fossil record and embryological studies (Romer and Parsons, 1986). However, the simpler the morphology of an organism, the harder it is to trace the origins of its novel structures. For example, one of the most striking evolutionary transformations among animals was the evolution of an **amphioxus**-like animal (Figure 1.17) into a fish. Amphioxus lacks most of the structures found in the heads of all vertebrates—it does not have an organized brain, it lacks organized sensory structures such as eyes, and it has no skull. How did the vertebrate head, with its well-developed brain and sense organs originate from such humble beginnings? Is the vertebrate head a new feature, added to the anterior end of an amphioxus-like ancestor, or is the vertebrate head just a highly modified front end of an amphioxus-like animal? Proposal of the first of these ideas, the "new head" hypothesis (Gans and Northcutt, 1983; Northcutt and Gans, 1983; Gans, 1989), was prompted in part by the observation that the **notochord**, a stiff, phosphatic rod that functions to preserve body shape during locomotion in amphioxus, runs the entire length of the body in amphioxus,

FIGURE 1.16

Evolutionary transformations between species. *Part A*: All pairs of organisms share a common ancestor. However, dinosaurs such as *Velociraptor* (from Paul, 1989; reprinted with permission) did not evolve from arthropods such as the amphipod crustacean *Cyamus* that is parasitic on whales (from Brusca and Brusca, 1990; reprinted with permission), or vice versa. *Part B*: Central questions in evolutionary biology are what did the common ancestors of organisms such as *Velociraptor* and *Cyamus* look like, and how did they evolve to become the types of organisms known in the fossil record or in the modern biota?

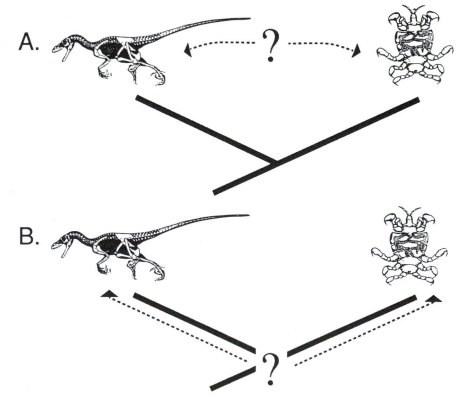

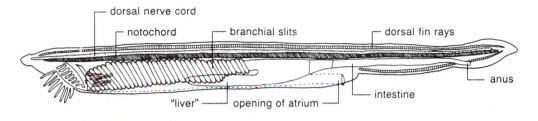

dorsal nerve cord
notochord
branchial slits
dorsal fin rays
anus
intestine
"liver"
opening of atrium

FIGURE 1.17

Vertebrate animals evolved from an amphioxus-like ancestor. Drawing shows some internal structures of the amphioxus *Branchiostoma*. Note that the notochord runs to the tip of the snout, and that there are no organized sense organs or a brain. The ancestors of fish and all other vertebrates may have looked somewhat like an amphioxus. An adult amphioxus seldom exceeds five centimeters in length. (Figure from *The Vertebrate Body*, Sixth Edition by Alfred S. Romer and Thomas S. Parsons, copyright © 1986 by Saunders College Publishing, reproduced by permission of the publisher.)

but extends only as far forward as the midbrain in the embryos of vertebrates. Although intriguing, this hypothesis is difficult to test.

In this case, our understanding of the genetic basis of development offers great potential. A series of genes, called **homeobox** genes (discussed in Chapter 5), are crucial to development. Many of the homeobox genes found in the fruitfly, *Drosophila*, are also found in amphioxus and in vertebrates. (These genes are **homologous**, i.e., they share a common evolutionary derivation.) Remarkably, many of these homologous genes have been found to be expressed in the same part of the body in very different animals. For example, homologous homeobox genes in amphioxus and chicken are found to affect essentially the same positions along the head to tail (anteroposterior) axis of both organisms (Holland et al., 1992). This suggests the new head hypothesis is incorrect (Peterson, 1994). Similarly, genes expressed in the developing brains of *Drosophila* are also found in the developing brains of mice (Simeone et al., 1992), which suggests that similarities between vertebrate and arthropod heads are not merely a matter of coincidence, but remarkably are due to the same developmental mechanism. In short, both dinosaurs and arthropods (Figure 1.16B), despite their dramatic differences, share essentially the same sort of developmental program for the head, one that was presumably present in their last common ancestor.

The studies by Holland et al. (1992) and Simeone et al. (1992) show that developmental biology offers great potential to help explain the origin of new structures; patterns of gene expression can be used to identify homologous regions of the body long after those regions have become so different in form and shape as to be otherwise unrecognizably related.

EVOLUTION AS THE CONTROL OF DEVELOPMENT BY ECOLOGY

The simple two-step view of the evolutionary process—variation followed by selection—does not do justice to the complexity of the process as it occurs in nature. However, there is a way to consider the two-step process to address some of the underlying complexity by invoking two central subdisciplines of biology: Developmental biology and ecology. Variation in the phenotype among members of a species' population is a result of development. Natural selection is a consequence of the interactions of organisms and their environments; thus, natural selection lies primarily in the domain of ecology. Hence, the two-step characterization of the Darwinian syllogism, variation followed by selection, can be re-expressed as the control of development by ecology (Van Valen, 1973, 1974).

This characterization is not quite satisfactory, in that ecology does not exactly control development. Rather, development produces forms and ecological interactions help determine which of these forms generated by developmental processes survives in the next generation. Nonetheless, this recasting of variation-followed-by-selection focuses attention on the interrelations between two central aspects of biology: development and ecology.

Pioneering developmental biologists, such as Ernst Heinrich Haeckel (1834–1919) and Karl Ernst von Baer (1792–1876), were major contributors to evolutionary biology in the late nineteenth and early twentieth centuries. In the last few decades molecular techniques have revitalized the study of development, yet developmental biology has made relatively little contribution to evolutionary studies during this time. The source and nature of variation have received only limited attention (Raff and Kaufman, 1983), perhaps because of the inherent difficulties in analyzing organismal development. In the future there should be a much greater contribution of developmental biology to our understanding of evolution. Some of the most exciting developments in the field are surveyed by De Robertis in Chapter 5 and in Akam et al. (1994).

THE UNEXPECTED: DNA AS THE NOT-SO-STATIC BLUEPRINT

DNA is commonly regarded as the "blueprint of life," and as such may be viewed as a passive archive of information, sequestered away from the normal dynamics of cellular processes, although it participates in cell division, gene expression, and the like. However, the organization of the DNA itself may be modified by cellular processes (e.g., as discussed above regarding the incorporation of the 10-kb retrovirus DNA from primates into the cat genome). Another interesting example is the case where manipulation of the DNA is needed to activate genes that are required for major morphological and biochemical changes in certain types of eubacterial **cyanobacteria**. Nitrogen is necessary for life; it is required for the synthesis of **amino acids**, the building blocks of proteins, and **nucleotides**, the building blocks of DNA and RNA (discussed in Chapter 4). The most abundant source of nitrogen is atmospheric nitrogen, N_2. However, only a relatively few species of microorganisms can make use of (fix) N_2 directly. One of the most diverse groups of nitrogen-fixing organisms are the cyanobacteria.

Filamentous cyanobacteria consist of strings of cells that do not fix N_2 if other sources of nitrogen (such as ammonia or nitrate) are available. However, if starved of these alternate nitrogen sources, certain cyanobacteria develop specialized cells called **heterocysts** at regular intervals along the filament (Figure 1.18). Within each heterocyst, approximately 1000 heterocyst-specific gene products are synthesized that, among other things, build the machinery needed to fix nitrogen from the atmosphere (Haselkorn, 1992). The thick wall of the heterocyst is important in keeping out oxygen, which inhibits nitrogen fixation.

Among the more important genes in nitrogen fixation are the *ni*trogen *f*ixation genes themselves, the **nif** genes. In vegetative (normal) cells of the cyanobacterium *Anabaena*, a crucial *nif* gene, *nifD,* and another gene, *fdxN,* are each split into two segments, sepa-

FIGURE 1.18

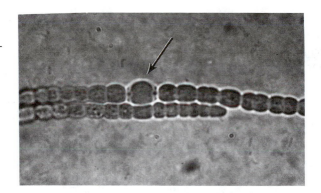

Two filaments of the heterocystous cyanobacterium *Anabaena*. The large cell in the upper filament (arrow) is a heterocyst, a cell specialized for nitrogen fixation. (Photograph provided by J. William Schopf, University of California, Los Angeles.)

rated by extremely large stretches of non-coding DNA (**insertions**). Functional proteins cannot be made from either of these two genes with the insertions present. However, as part of the process of heterocyst formation, the intervening pieces of DNA are cut out of the genome, reuniting the severed regions of each of the genes, and nitrogen fixation can proceed (Figure 1.19). Interestingly, the enzymes required to excise the insertions are coded for within the insertions themselves (the *xisA* and *xisF* genes, stippled in Figure 1.19). Thus, in *Anabaena*, the DNA itself is modified as part of heterocyst formation.

Another example involves "mobile DNA," which has been implicated in human disease. In the human genome there is a sequence of approximately 300 nucleotides, called the "Alu sequence," of which more than 500,000 copies occur in each cell. It is unclear

Anabaena Vegetative Cell

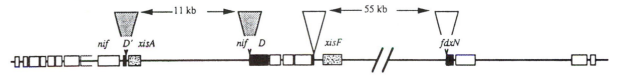

Anabaena Heterocyst

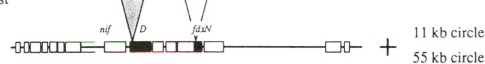

11 kb circle

55 kb circle

FIGURE 1.19

Genomic region shown containing some genes crucial to formation and function of heterocysts of cyanobacterium *Anabaena*. Genes are represented by boxes. Stretches of DNA not coding for protein are indicated by the solid lines. In vegetative cells (those unable to fix nitrogen), two of the heterocyst-specific genes are interrupted by large insertions of DNA, one 11 kb, the other 55 kb (upper panel). In cells that become heterocysts, these two insertions are excised, producing two nonfunctional circles of DNA, which unite the separated portions of the *nifD* and *fdxN* genes and enable them to function (lower panel). Each insert is excised with the aid of an enzyme coded for by a gene that lies within the insert. The 11-kb insertion that disrupts the *nifD* gene is released with the enzyme coded by the *xisA* gene (stippled, upper panel); the 55-kb insert that disrupts the *fdxN* gene is removed as a result of the activity of the *xisF* gene (stippled, upper panel). (Modified from Haselkorn, 1992. Reproduced, with permission, from the Annual Review of Genetics Volume 26, © 1992 by Annual Reviews Inc.)

whether any of these copies serves any function. From time to time, new copies of Alu are produced and inserted into the genome. Recently, Goldberg et al. (1993) proposed that some patients with Huntington's disease—a fatal disease that appears in mid-life, causing personality and cognitive disturbances and involuntary movements—may be caused by the insertion of an Alu sequence into a sensitive region of the genome.

Our knowledge of the human genome and of its properties is far from complete. New discoveries will not only increase our understanding of the evolution of life, but may also offer new approaches and possibilities for control of disease. In Chapter 6, Simpson reviews the progress that has been made in developing diagnoses for one of the most serious diseases in South and Central America, Chagas' disease. The tools being developed are centered around the very unusual and remarkable way that some proteins are encoded in the genome of the parasites that cause Chagas' disease. We will increasingly find surprises as we continue to probe the genomes of the natural world.

CONCLUSION

Darwin established beyond a reasonable doubt that natural processes are responsible for the origin and transformation of biological species through time. His central contribution was a syllogism, a logical construct, that provided a mechanism for the occurrence of descent with modification. The basic syllogism remains sound, though there have been elaborations and changes in emphasis as explanations for phenotypic evolution and genotypic evolution have been propounded. In Darwin's day, many basic properties of biological systems were a mystery. Although many still remain mysterious, the molecular revolution has enabled questions to be answered that were wholly unanswerable in Darwin's time, and has opened many new areas of evolutionary inquiry. The five chapters that follow provide a more comprehensive introduction to some of the central breakthroughs contributed by the molecular revolution to studies of the evolution of life.

Acknowledgments

I thank John Bragin, Cliff Brunk, Ernst Mayr, Bill Schopf, and Bruce Weber for their helpful comments and suggestions, and Richard Mantonya for assistance with the figures.

REFERENCES

Akam, M., Holland, P., Ingham, P., and Wray, G. 1994. *The Evolution of Developmental Mechanisms.* Development 1994 Supplement (Cambridge, U.K.: Company of Biologists).

Allison, A.C. 1956. Sickle cells and evolution. *Scientific American 195(2):* 87–94.

Alvarez, L.W. 1983. Experimental evidence that an asteroid impact led to the extinction of many species 65 million years ago. *Proc. Natl. Acad. Sci. USA. 80:* 627–642.

Alvarez, L.W., Alvarez, W., Asaro, F., and Michel, H.V. 1980. Extraterrestrial cause for the Cretaceous-Tertiary extinction. *Science 208:* 1095–1108.

American Geophysical Union Special Report on Volcanism and Climate Change. May, 1992 (Washington, D.C.: American Geophysical Union).

Avise, J.C., and Nelson, W.S. 1989. Molecular genetic relationships of the extinct Dusky Seaside Sparrow. *Science 243:* 646–648.

Benton, M.J. 1990. *Vertebrate Paleontology* (London: Unwin Hyman).

Benveniste, R. 1985. The contributions of retroviruses to the study of mammalian evolution. In: R. J. MacIntyre (Ed.), *Molecular Evolutionary Genetics* (New York: Plenum), pp. 359–417.

Bowen, B., Avise, J.C., Richardson, J.I., Meylan, A.B., Margaritoulis, D., and Hopkins-Murphy, S.R. 1993. Population structure of loggerhead turtles (*Caretta caretta*) in the Northwestern Atlantic Ocean and Mediterranean Sea. *Conserv. Biol. 7:* 834–844.

Brunk, C.F. 1991. Darwin in an age of molecular revolution. *Contention 1:* 131–150.

Brusca, R.C. and Brusca G.J. 1990. *Invertebrates* (Sunderland, MA: Sinauer).

Colbert, E.H. and Morales, M. 1991. *Evolution of the Vertebrates* (4th ed.) (New York: Wiley-Liss).

Coyne, J.A. 1994. Ernst Mayr and the origin of species. *Evolution 48:* 19–30.

Darwin, C. 1859. *On the Origin of Species.* (1st ed.) (London: John Murray).

De Rijk, P., Neefs, J.-M., Van de Peer, Y., and De Wachter, R. 1992. Compilation of small ribosomal subunit RNA sequences. *Nuc. Acid Res. 20:* 2075–2089.

Frolich, L.M. 1991. Osteological conservatism and developmental constraint in the polymorphic "ring species" *Ensatina eschscholtzii* (Amphibia: Plethodontidae). *Biol. J. Linn. Soc. 43:* 81–100.

Gans, C. 1989. Stages in the origins of vertebrates: Analysis by the means of scenarios. *Biol. Rev. Camb. Phil. Soc. 64:* 221–268.

Gans, C., and Northcutt R.G. 1993. Neural crest and the origin of vertebrates: A new head. *Science 220:* 268–274.

Goldberg, Y.P., Rommens, J.M., Andrew, S.E., Hutchinson, G.B., Lin, B., Theilmann, J., Graham, R., Glaves, M.L., Starr, E., McDonald, H., Nasir, J., Schappert, K., Kalchman, M.A., Clarke, L.A., and Hayden, M.R. 1993. Identification of an Alu retrotransposition event in close proximity to a strong candidate gene for Huntington's disease. *Nature 362:* 370–373.

Gould, S.J. 1980. *The Panda's Thumb* (New York: Norton).

Gould, S.J. 1989. *Wonderful Life* (New York: Norton).

Haselkorn, R. 1992. Developmentally regulated gene rearrangements in prokaryotes. *Annu. Rev. Genet.* 26: 113–130.

Hedges, S.B., Moberg, K.D., and Maxson, L.R. 1990. Tetrapod phylogeny inferred from 18S and 28S ribosomal RNA sequences and a review of the evidence for amniote relationships. *Mol. Biol. Evol. 7:* 607–633.

Holland, P.W.H., Holland, L.Z., Williams, N.A., and Holland, N.D. 1992. An amphioxus homeobox gene: Sequence conservation, spatial expression during development and insights into vertebrate evolution. *Development 116:* 653–661.

Kimura, M. 1968. Evolutionary rate at the molecular level. *Nature 217:* 624–626.

King, L.K., and Jukes, T.H. 1969. Non-Darwinian evolution. *Science 164:* 788–798.

Lewontin, R.C. 1983. The organism as the subject and object of evolution. *Scientia 118:* 65–82.

Mayr, E. 1942. *Systematics and the Origin of Species* (New York: Columbia Univ.).

Mayr, E. 1988. *Toward a New Philosophy of Biology* (Cambridge: Harvard Univ. Press).

Mayr, E. 1991. *One Long Argument* (Cambridge: Harvard Univ. Press).

Moritz, C., Schneider, C.J. and Wake, D.B. 1992. Evolutionary relationships within the *Ensatina eschscholtzii* complex confirm the ring species interpretation. *Syst. Biol. 41:* 273–291.

Niklas, K.J. 1994. Morphological evolution through complex domains of fitness. *Proc. Natl. Acad. Sci. USA 91:* 6772–6779.

Northcutt, R.G., and Gans, C. 1983. The genesis of neural crest and epidermal placodes: A reinterpretation of vertebrate origins. *Quart. Rev. Biol. 58:* 1–28.

Olsen, G.J. 1987. Earliest phylogenetic branchings: Comparing rRNA-based evolutionary trees inferred with various techniques. *Cold Spring Harbor Symp. Quant. Biol. 52:* 825–837.

Paul, G.S. 1989. *Predatory Dinosaurs of the World* (New York: Simon & Schuster).

Peterson, K.J. 1994. The origin and early evolution of the Craniata. *In* D.R. Prothero and R.M. Schoch (eds.), *Major Features of Vertebrate Evolution.* Short Courses in Paleontology 7: 14–37.

Raff, R.A., and Kaufman, T.C. 1983. *Embryos, Genes and Evolution* (New York: Macmillan).

Raup, D.M. 1991. *Extinction: Bad Genes or Bad Luck?* (New York: Norton).

Ridley, M. 1993. *Evolution* (Boston: Blackwell Scientific).

Romer, A.S., and Parsons, T.S. 1986. *The Vertebrate Body* (6th ed.). (Philadelphia: Saunders).

Sharpton, V.L., Burke, K., Camargo-Zanoguera, A., Hall, S.A., Lee D.S., Marín, L.E., Suárez-Reynoso, G., Quezada-Muneton, J.M., Spudis, P.D., and Urrutia-Fucugauchi, J. 1993. Chicxulub multiring impact basin: Size and other characteristics derived from gravity analysis. *Science 261:* 1564–1567.

Simeone, A., Acampora, D., Gulisano, M., Stornaiuolo, A., and Boncinelli, E. 1992. Nested expression domains of four homeobox genes in developing rostral brain. *Nature 358:* 687–690.

Simpson, G.G. 1964. Organisms and molecules in evolution. *Science 146:* 1535–1538.

Van Valen, L. 1974. A natural model for the origin of some higher taxa. *J. Herp. 8:* 109–121.

Van Valen, L. 1973. Festschrift. *Science 180:* 488.

Van Valkenburgh, B., and Hertel, F. 1993. Tough times at La Brea: Tooth breakage in large carnivores of the late Pleistocene. *Science 261:* 456-459.

Wayne, R.K. 1986. Cranial morphology of domestic and wild canids: The influence of development on morphological change. *Evolution 40:* 243–261.

Weiner, J. 1994. *The Beak of the Finch* (New York: Knopf).

■

FURTHER READING

Darwin and Darwinism

Darwin, C. 1859. *On the Origin of Species* (Facsimilie of 1st ed.). Reprinted 1979 (Cambridge: Harvard Univ. Press).

Dawkins, R. 1987. *The Blind Watchmaker: Why the Evidence of Evolution Reveals a Universe Without Design* (New York: Norton).

Depew, D.J., and Weber, B.H. 1995. *Darwinism Evolving* (Cambridge: MIT Press).

Ghiselin, M.T. 1969. *The Triumph of the Darwinian Method* (Berkeley: Univ. California Press).

Mayr, E. 1991. *One Long Argument* (Cambridge: Harvard Univ. Press).

Weiner, J. 1994. *The Beak of the Finch* (New York: Knopf).

Textbooks on Evolution, Development and Evolution, and Molecular Biology

Futuyma, D.J. 1986. *Evolutionary Biology* (Sunderland, MA: Sinauer).

Raff, R.A., and Kaufman, T.C. 1983. *Embryos, Genes and Evolution* (New York: Macmillan).

Ridley, M. 1993. *Evolution* (Boston: Blackwell Scientific).

Skelton, P. 1993. *Evolution: A Biological and Paleontological Approach* (Reading, MA: Addison-Wesley).

Watson, J.D., Hopkins, N.H., Roberts, J.W., Steitz, J.A., and Weiner, A.M. 1987. *Molecular Biology of the Gene* (Menlo Park, CA: Benjamin/Cummings).

CHAPTER 2

HOW DID THE MOLECULAR REVOLUTION START? WHAT MAKES EVOLUTION HAPPEN?

∎

Thomas H. Jukes*

∎

INTRODUCTION

During the past 30 years, the study of evolution changed from an organismal science to a molecular science. Classic evolutionists, starting with Darwin, had studied and compared the anatomy and morphology of both living and fossil organisms. Earlier, the Swedish systematist, Carolus Linnaeus (1707–1778), had classified species on the basis of the similarities of their morphological, visible characteristics—the similarities of their **phenotypes**. Darwin saw that descent with modification had taken place, and that natural selection determined which species would survive the struggle for existence. His scientific successors discovered mutations (a change in a gene) and showed that mutations provided variations of phenotypic characters that were the raw material for natural selection. The units of heredity were found to be located in chromosomes, the sites of the genetic information (known collectively as the **genotype**), and hereditary units could be mapped into their individual **genes**, each of which was shown to govern a specific trait.

Biochemistry entered the scene in 1910, when Archibald Garrod discovered that certain biochemical peculiarities in humans, such as alcaptonuria, in which the urine of the subject turns black on standing, could be inherited. He called these peculiarities "inborn errors of metabolism." Because metabolism is controlled by **enzymes**, proteins that mediate the metabolic chemical reactions (see Chapter 4), attention focused on enzymes as the expression of genes. In 1936, Beadle and Tatum, on the basis of studies of the bread mold *Neurospora*, proposed a **one gene–one enzyme theory of inheritance**: Because a mutation was observed to inactivate a corresponding specific enzyme, each gene must control the formation of a single enzyme. Molecular changes in evolution can be adaptive, neutral, or deleterious (and sometimes lethal). The changes originate from errors occurring in the chromosomes during DNA replication and gene duplication. In 1952, Sanger showed that in insulin, the first protein to be sequenced, single amino acid differences among humans, horses, pigs, and whales existed, but that these differences made no difference to the clinical effectiveness of insulin. These amino acid differences, therefore, were "neutral" changes. In 1957, Vernon Ingram found that replacement of a single amino acid in a mutated hemoglobin molecule produced

*Department of Integrative Biology, University of California, Berkeley, CA 94720.

the hereditary disease sickle cell anemia (discussed in Chapter 1). This remarkable discovery brought **protein** molecules, each composed of a linear array of amino acids, into genetics, and was the starting gun in the race to determine the sequence of amino acids in proteins and to compare **homologous** proteins, those sharing a common protein evolutionary ancestor, in different species of organisms.

Classical evolutionary theory concerns mainly the behavior in populations and the ecology of organismal phenotypes. The phenotype is the result of action of several types of genes, not only of **structural genes**, those that result in production of proteins, but also of genes that regulate biochemistry and that dictate the cellular development of organisms. Thus, the phenotype results from a complex series of events, and because of this, its characteristics cannot be predicted from its genome.

Molecular evolutionary theory tends to be **reductionist**, in contrast to that of classic evolution. Reductionism addresses scientific questions by reducing complex data or phenomena to simple terms. In essence, reductionism is simplification, and once simplification has been achieved, more complicated constructs can be built by use of basic rules. An outstanding example of the use of reductionism to dissect and comprehend a complex system is shown by our understanding of the DNA molecule, which reduces genetics and evolution to the bare bones of the very specific rules of **base pairing**, the formation in DNA of **hydrogen bonds** that link together the nitrogen-containing chemical bases **adenine** and **thymine** (A and T) and the bases **cytosine** and **guanine** (C and G), as shown in Figure 2.1.

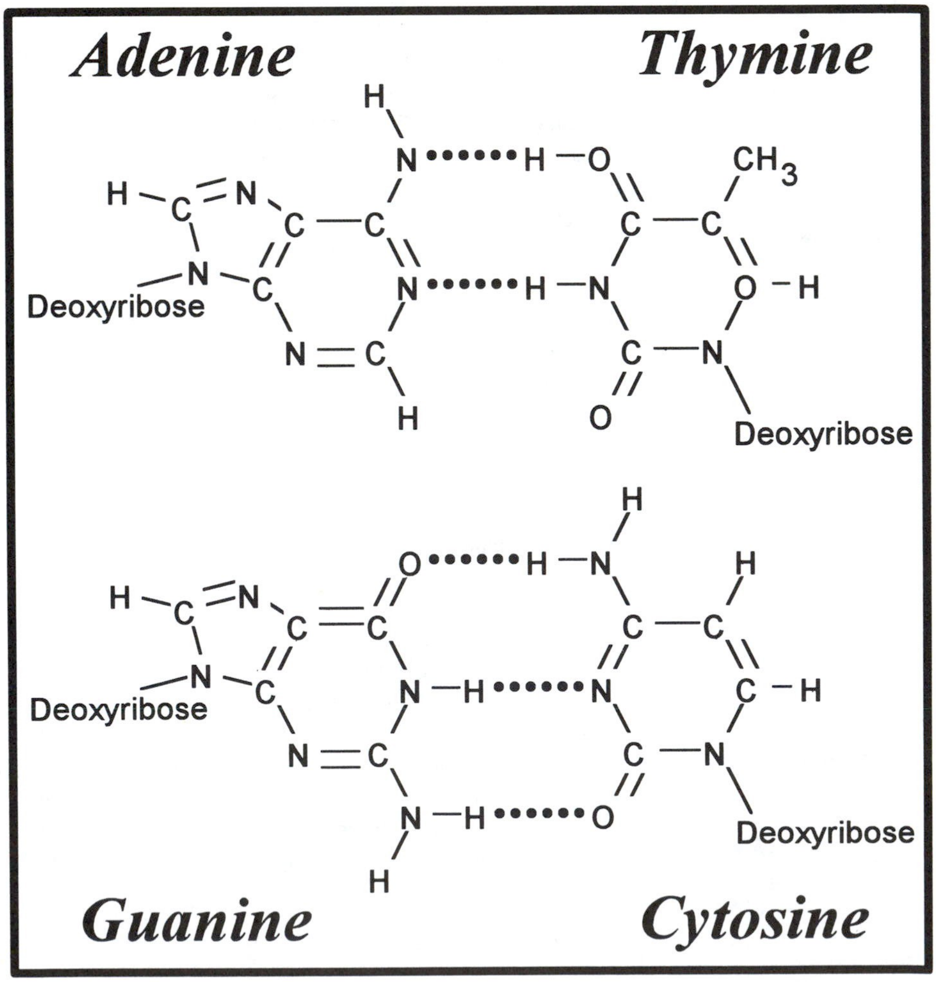

FIGURE 2.1

The standard base pairs in DNA, linked together by hydrogen bonds (dotted lines). When the information in DNA (specified by the order of the DNA bases) is transcribed into mRNA, uracil in the RNA forms hydrogen bonds with adenine.

Norman Horowitz (1994) recently wrote [emphasis added]

> Antireductionism is not a scientific position, but a political one . . . [antireductionism] is actually antiscience. . . . Science *must* be reductionist . . . natural systems can be said to be understood only after they have been reduced to and reassembled from their components.

There is no obvious reason why proteins, such as blood hemoglobins or respiratory cytochromes *c*, should not be identical in all mammals. After all, all animals use identical protein-associated coenzymes, such as riboflavin, thiamine, and nicotinamide adenine dinucleotide (NAD). But when the molecules with the same function are proteins, differences can be found between species. Realization of this fact led to the start of a new branch of science, **molecular evolution**.

Studies of molecular evolution serve to expand and quantify those of the classical evolutionists. An early, powerful finding was that there is about 60% common identity between the amino acid sequences of the respiratory cytochrome *c* proteins in humans and yeast; yet, obviously, humans and yeast are certainly not closely related. This degree of molecular similarity was therefore unexpected—a startling result, and one not predicted by classical evolutionary theory.

Given today's knowledge, evolution cannot be presented adequately without the foundation provided by molecular biology. Much has been made of gaps in the fossil record, but there is no such problem when comparisons are made in living organisms of the base sequences of 16S/18S ribosomal RNA molecules (discussed in Chapters 1, 3, and 4) or of hemoglobin genes.

To accommodate the accumulating data and insights in this new field, many scientific journals have been started, for example, the *Journal of Molecular Evolution* (1971), *Molecular Biology and Evolution* (1983), and *Molecular Phylogenetics and Evolution* (1992). Numerous articles on molecular evolution have appeared in leading general scientific journals that publish results of biologic studies, such as *Nature* and the *Proceedings of the National Academy of Sciences*. This chapter discusses two main components of molecular evolution: mutation and selection.

THE ROLE OF DNA

The Genetic Code and the Synthesis of Proteins

Proteins (such as the globins of blood and cytochrome *c* molecules) are composed of a linear array of amino acid subunits, typically 100 or more amino acids in length. Twenty different amino acids are used in protein synthesis, and it is the order in which they are arranged in a protein molecule that determines its function in an organism. This order, in turn, is specified by the ordered arrangement of the four nitrogen-containing bases (A, T, G, C) present in the **DNA** (deoxyribonucleic acid) molecules of the **chromosomes**. The information-containing molecule, DNA, is in the chromosomes, but proteins are made elsewhere in the cell, on small, protein-manufacturing bodies called **ribosomes**. The information in the DNA of chromosomes needed to synthesize proteins is carried to the ribosomes by molecules of **mRNA** (messenger ribonucleic acid) that are similar to DNA but that, in addition to a few other changes, have the nitrogenous base uracil (U) in place of thymine (T).

The chemical structure of the four bases in DNA, and the way that they base pair, are shown in Figure 2.1. The two standard base pairs are A, adenine, with T, thymine (or, A of the DNA with U, uracil, of the RNA), and G, guanine, with C, cytosine. G also some-

FIGURE 2.2

The universal amino acid code in mRNA that results in insertion of the amino acids indicated into proteins produced at ribosomes.

		SECOND BASE OF RNA CODE				
		URACIL	CYTOSINE	ADENINE	GUANINE	
FIRST BASE OF RNA CODE	**URACIL**	UUU (Phe) UUC (Phe) UUA (Leu) UUG (Leu)	UCU (Ser) UCC (Ser) UCA (Ser) UCG (Ser)	UAU (Tyr) UAC (Tyr) UAA (Stop) UAG (Stop)	UGU (Cys) UGC (Cys) UGA (Stop) UGG (Trp)	U C A G
	CYTOSINE	CUU (Leu) CUC (Leu) CUA (Leu) CUG (Leu)	CCU (Pro) CCC (Pro) CCA (Pro) CCG (Pro)	CAU (His) CAC (His) CAA (Gln) CAG (Gln)	CGU (Arg) CGC (Arg) CGA (Arg) CGG (Arg)	U C A G
	ADENINE	AUU (Ile) AUC (Ile) AUA (Ile) AUG (Met)	ACU (Thr) ACC (Thr) ACA (Thr) ACG (Thr)	AAU (Asn) AAC (Asn) AAA (Lys) AAG (Lys)	AGU (Ser) AGC (Ser) AGA (Arg) AGG (Arg)	U C A G
	GUANINE	GUU (Val) GUC (Val) GUA (Val) GUG (Val)	GCU (Ala) GCC (Ala) GCA (Ala) GCG (Ala)	GAU (Asp) GAC (Asp) GAA (Glu) GAG (Glu)	GGU (Gly) GGC (Gly) GGA (Gly) GGG (Gly)	U C A G

(right side label: **THIRD BASE OF RNA CODE**)

ABBREVIATIONS

THREE LETTER	SINGLE LETTER	AMINO ACID
Phe	F	Phenylalanine
Leu	L	Leucine
Ile	I	Isoleucine
Met	M	Methionine
Val	V	Valine
Ser	S	Serine
Pro	P	Proline
Thr	T	Threonine
Ala	A	Alanine
Tyr	Y	Tyrosine

ABBREVIATIONS

THREE LETTER	SINGLE LETTER	AMINO ACID
His	H	Histidine
Gln	Q	Glutamine
Asn	N	Asparagine
Lys	K	Lysine
Asp	D	Aspartic acid
Glu	E	Glutamic acid
Cys	C	Cysteine
Trp	W	Tryptophan
Arg	R	Arginine
Gly	G	Glycine

times pairs with U; a G-U pair plays an important role in **translation** of the genetic code, the process that determines which of various amino acids are inserted into proteins.

There are many different amino acids in living systems, but only 20—the *magic 20*—take part in the biological synthesis of proteins. These 20 are listed in Figure 2.2, the universal amino acid code, together with the three-letter and single-letter abbreviations of the amino acid names. In addition to these 20, other amino acids occur in some proteins, but these are produced by chemical modification of one or another of the magic 20 after protein synthesis has been completed. For example, the amino acid hydroxyproline is abundant in collagen, the fibrous protein of vertebrate bones, but it is formed by the addition of a hydroxyl (-OH) group to proline, one of the magic 20, after the protein has been synthesized.

The first step in the series of events that leads to the synthesis of a protein involves transfer of the information contained in DNA (information specified by the order of the

DNA bases) to a molecule of mRNA, a process known as **transcription**. To accomplish transcription, the double-stranded DNA "unzips" the hydrogen bonds linking together the base pairs of the two strands—that is, the bonds are temporarily broken—and a single-stranded mRNA molecule is formed alongside one of the two strands. For example, let us imagine that the base sequence in the two strands of a molecule of DNA, held together by hydrogen bonding of base pairs (shown by dots), is:

DNA Strand 1: A-T-C-C-G-T-A-G-G-A-T-T

. . . .

. . . .

. . . .

DNA Strand 2: T-A-G-G-C-A-T-C-C-T-A-A

then, after the DNA unzips, the base sequence in DNA Strand 2 can be transcribed to a molecule of mRNA as follows:

C-C-G-T-A-G-G-A-T-T

/

DNA Strand 1: A-T

. .

mRNA: . . C-C-G-U-A-G-G-A-U-U

. . . .

DNA Strand 2: T-A . . . .

\ . . . .

G-G-C-A-T-C-C-T-A-A

Note that uracil (U) of the mRNA pairs with adenine (A) of the DNA; U in the RNA is thus the **transcript** of A in the DNA.

In DNA, each base is combined with phosphate and a particular sugar, deoxyribose. In RNA, the deoxyribose is replaced by a different sugar, ribose. In both these **nucleic acids**, the subunit composed of a base with a phosphate and a sugar is called a **nucleotide**.

The information transfer from chromosomal DNA to mRNA to the protein is accomplished by means of a three letter **genetic code**—each triplet of bases in one strand of the DNA (for example, cytosine-thymine-adenine, abbreviated C-T-A) is transcribed, by the rules of base pairing, into a corresponding triplet in the mRNA (in this example, G-A-U) known as the **codon**; and at the ribosome, the codon in the mRNA directs a specific amino acid (aspartic acid, in this example) to be inserted at the corresponding position in the manufactured protein molecule. Thus, a sequence of 300 bases in the DNA would be transcribed into a corresponding 300 base-long sequence in the mRNA, and this, in turn, would code for 100 amino acids in the protein product.

By the mRNA codes shown in Figure 2.2, each of 61 chromosomal DNA triplets codes for a specific amino acid. There are also three "stop" codons that terminate the synthesis of a protein, once its manufacture has been completed. Hence, changes in any of the bases in a DNA triplet—changes called **mutations**—can alter a codon. Thus, if the third DNA base in the above example were to be mutated from adenine to cytosine, the DNA code would be changed from C-T-A to C-T-C, the corresponding mRNA codon from G-A-U to G-A-G, and the amino acid inserted into the protein would be changed from aspartic acid to glutamic acid. Note, however, that because the genetic code is **redundant**—that is, because all but two of the 20 amino acids are specified by more than one codon—not all mutations cause amino acid changes. For example, if the third DNA base in the above triplet were mutated to guanine rather than to cytosine, the DNA code would be changed from C-T-A to C-T-G, and the mRNA codon from G-A-U to G-A-C, but aspartic acid would still be inserted at the same position in the protein because G-A-U and G-A-C both code for this same amino acid. Mutations such as this—because they do not cause changes in the amino acid sequence and thus are not detectable in the protein product—are known as **silent,** or **synonymous, mutations**.

The Genetic Code and Protein Evolution

In 1963, investigation of the redundancy of the genetic code had progressed to a stage at which it had become possible to distinguish between pairs of amino acids as being related by single-base or two-base changes in their codes (Jukes, 1963). Thus, for example, as shown in Figure 2.2, the four codes for valine and the four codes for alanine are separated by a single-base change (at the second of the three bases), and the codes for valine and the four codes for threonine by a change in two bases (the first two bases of the triplet). The two-base change results from two successive substitutions and, therefore, two successive amino acid replacements. For threonine to be replaced by valine, for example, threonine must first be replaced by either alanine or isoleucine, from a single-base change, and either of these amino acids must then be replaced by valine, from a second single-base change.

Aligning and comparing the amino acid sequences of two human hemoglobins (known as human alpha and gamma hemoglobins), shows that they are related by 55 single-base changes and 25 two-base changes that had occurred in 141 codons (Jukes, 1963). Averaging these changes (excluding synonymous changes) over the length of the molecule shows that the minimum number of base changes per codon is 0.74. This technique of considering both single-base and two-base changes, new in 1963, was an improvement over earlier approaches that considered only the total amino acid differences per codon (in this example, 0.56), because it more closely approximated the actual changes that had occurred in the DNA (Jukes, 1963). Later, when methods progressed so that it became possible to determine the actual sequences of nucleotides in the genes coding for such proteins, comparisons could be made in terms both of replacements *and* silent nucleotide changes per codon.

In 1966, Jukes pointed out [emphasis added]

> If we consider as an example the two cytochrome *c* molecules found respectively in dogs and horses, these differ in about six of the amino acids in a chain of 104. Are the two molecules tailored to different requirements, have they evolved to conform to these two different requirements, or have the two cytochromes passively been carried along as dogs and horses evolved separately from a common ancestor? In the latter case, it is undoubtedly quite probable that separation of the two species would be followed by changes in the genes that would result in differences in the two cytochrome *c* molecules. We do not know whether a horse could get along just as well with dog cytochrome *c* as with its own. The alpha chains of human and gorilla hemoglobins differ only with respect to substitutions of glutamic acid by aspartic acid at site 23. It taxes the imagination to infer that the single amino acid difference has become fixed by natural selection. . . . The changes produced in proteins by mutations will in some cases destroy their es-

sential functions *but in other cases the change allows the protein molecule to continue to serve its purpose.*

This was the first statement of the **neutral theory of molecular evolution**, the theory that most mutations (such as nucleotide changes) in chromosomal DNA have no effect whatever on the organismal phenotype—they are neither advantageous nor disadvantageous, simply neutral.

The first milestone of the molecular revolution in evolution was the discovery of the structure of DNA by Watson and Crick (1953). This was recognized as showing the molecular basis of heredity, and the science of **molecular genetics**, leading to the manipulation of genomes in order to produce genetically new strains of organisms, soon started. But the key aspect of the DNA molecule for evolutionary studies is that its sequence history can be followed backward through time. A person's DNA is inherited from parents, theirs from grandparents, and so on, all the way back to the first DNA molecule. If this process is interrupted, a species becomes extinct. It was evident early in the molecular revolution that DNA held the key that would unlock the evolutionary past. For example, somewhere in evolutionary history there was a DNA molecule of the common ancestor of humans and apes. Further back, there must have been an ancestral DNA molecule of the creature from which all mammals descended. This was molecular confirmation of Darwin's theory of descent with modification of all of life from a single organism.

It is impossible to obtain DNA from organisms that lived millions of years ago (except in a few unusual specimens, such as insects trapped in amber, like those discussed in Chapters 3 and 6). And, in the 1950s, it was impossible to determine the exact sequence of nitrogenous bases in DNA (a problem solved finally in the 1970s). However, even if the DNA could not be sequenced, the amino acid sequences of homologous proteins from different organisms could be determined and compared, and their descent from a common ancestor could be deduced. One such finding, especially striking, was that the cytochromes *c* of humans and yeast are 60% *identical*. This is clear, quantitative, molecular proof of descent with variation from a common ancestor. Subsequent to this finding, hemoglobins from many vertebrates were sequenced and were found to be similar. By the early 1970s, **phylogenetic** studies (to determine the evolutionary relations among organisms) were becoming increasingly molecular, and the genealogical relationships thereby deduced fully supported those of classical evolutionary theory. The *Journal of Molecular Phylogenetics and Evolution* is specifically devoted to this area of study. Because there are thousands of proteins in each living organism, and millions of different species, it will be a long time before this journal runs out of subject matter!

MISTAKES IN DNA SELF-COPYING

Directional Mutation Pressure

Accurate replication of DNA is essential for the maintenance of a living system. In contrast, for evolution to take place, some inaccuracies, some mutations, must occur. There are genes called **mutator genes** that favor just such mistakes. Sometimes these genes bring about replacement of a G-C pair with an A-T pair, and at other times, the reverse occurs.

In 1961, Noboru Sueoka noted that there was a wide range in the G + C content of the DNA of bacteria. Sueoka investigated the amino acid composition of bacterial proteins and found that the GC content of DNA was related to the percentages of certain amino acids in the total protein. High GC content was correlated with relatively high amounts of the amino acids alanine, arginine, and glycine, and high AT content, with high isoleucine, lysine, phenylalanine, and tyrosine content. The genetic code, unknown in 1961, had been deciphered by 1966 and showed that the GC and AT contents of the first two base positions in the code explained Sueoka's results. In 1962 Sueoka had pointed out that when the GC contents of DNA in two organisms differed appreciably, enzymes of identical function had similar regions of primary function (**active sites**) but that "the dispensable parts of the molecule will be quite different." This assertion by Sueoka implied the occurrence of *neutral* amino acid changes in proteins of identical function, with such changes occurring in the "dispensable parts" of the molecule. At about the same time, Ernst Freese (1962) reached similar conclusions, stating that portions of proteins "can be partially altered without any functional change" as a result of changes in the base ratio; that "most base pairs in DNA can undergo changes that have no or only an insignificant selective effect"; and that "for each DNA species, one kind of base pair, e.g., GC, has been altered more frequently than the other one, resulting in a shift of the base ratio."

When the genetic code became known in the mid-1960s, it became possible to measure the frequency of mutational changes that result from **directional mutation pressure** affecting the silent third-base sites of codons (Figure 2.2). Jukes (1965) proposed that microorganisms high in G + C had probably "evolved towards the use of coding triplets ending in G or C" and that those low in G + C had "evolved towards the use of coding triplets ending in A or U." Cox and Yanofsky (1967) showed that the base ratio changed in the direction of increased G + C when the "Treffers mutator gene," which favors the substitution of A-T by C-G, was introduced into the bacterium *Escherichia coli*, and that the organism could flourish despite this change. Two years later, King and Jukes interpreted this finding as providing support for the neutral theory of molecular evolution (King and Jukes, 1969).

Some investigators claim that **synonymous nucleotide substitutions** (silent mutations of one base for another, for example, at most third-base sites of codons) are not truly neutral because they may affect the three-dimensional (secondary) molecular structure or other similar characteristics of mRNA (Mayr, 1993). This claim is countered, however, by the effect of directional mutation pressure on third-base codon positions in homologous genes. Jukes and Bhushan (1986) found that in the bacterium *Pseudomonas aeruginosa*, which has a G + C content of 68%, the GC content of silent (e.g., third-base site) positions in the gene coding for the enzyme tryptophan synthase is 92%, whereas it is only 40% for the same enzyme in the bacterium *Bacillus subtilis*, which has a G + C content of 45%. Thus, evidently, there is practically no evolutionary resistance to synonymous substitutions, and they are therefore neutral or very nearly neutral.

Jukes and Bhushan (1986) showed that in four different species of bacteria, the genes for the tryptophan synthase A and B enzymes have codon changes that have resulted in increased AT or GC content of their DNA, leading to effects in amino acid content corresponding to the changes noted by Sueoka (1961). This was the first direct verification of Sueoka's proposal. Jukes and Bhushan also measured the silent changes in the same series of codons, and found that these occurred nearly seven times as often as the changes that resulted in amino acid replacements. In this set of homologous genes, directional mutation pressure was therefore about seven times as effective in changing silent sites as in changing coding sites. A similar ratio of changes in silent sites compared with coding sites was found in mitochondrial genes of one insect and four vertebrate animals. When the GC content of such genes is known in many species, these observations should be able to be extended to other sets of homologous genes. However, if the GC content of the genes compared is constant, mutational changes will have taken place, but by a directionless process known as **genetic drift**. Directional mutation pressure, in contrast, channels the direction of the changes.

Histones

The nucleotide sequences of the genes for several chromosome-associated proteins (**histones**) in the sea urchins *Strongylocentrotus purpuratus* and *Psammechinus miliaris* have been reported by Schaffner et al. (1978) and Sures et al. (1978). Alignment and comparison of the genes for histone 3 in these two species showed 45 differences in the 402 pairs of nucleotides compared in the alignment occurring in 136 codons (three of which are incompletely sequenced in one of the species). None of the 45 nucleotides that had undergone substitution changed the amino acids inserted into the protein; these substitutions were therefore *not* related to protein function; that is, they are neutral.

What brought about these substitutions? There are two possibilities. First, by some unknown mechanism, the process of substitution recognized the silent nucleotide sites in codons, and selected them for substitution. This, however, seems impossible—the process of mutation affects chromosomal DNA, not the protein product, and occurs in noncoding as well as coding regions of DNA. There appears to be no way for the process of mutation to "know" which are the silent positions in the DNA triplets and which are not. The second explanation, which seems correct, is that mistakes are made in DNA replication, resulting in **point mutations** (changes of a single DNA base) that are subjected to natural selection. These mistakes are inherent in the replication process and are time-related. Point mutations that change the amino acids produced by the gene for histone 3 in sea urchins are deleterious or lethal, and hence are not accepted—that is, if they occur, the organism is at a disadvantage, or it dies, and the mutant gene is not carried over into subsequent generations. The 45 substitutions (occurring in 11% of the DNA base-containing nucleotides) are neutral changes that have accumulated since the divergence of the two sea urchin species from a common ancestor, probably about 50 million years ago.

When the genes for histone 3 in sea urchins are compared with those in rainbow trout (Connor et al., 1984; Winkfein et al., 1985), the number of silent substitutions rises to 61, reflecting the relatively greater evolutionary distance separating these two types of organisms, and there is one obviously unimportant amino acid replacement (Asp to Glu) at position 81 in the molecule. Thus, histones 3 show the "barrier effect" (discussed later) against amino acid replacements, and the ready acceptance of silent changes.

In 1971, Richard Dickerson published a diagram in the *Journal of Molecular Evolution* showing that each of several proteins had its own **unit evolutionary period**, defined as the length of time for a 1% change to occur in the amino acid sequence of the protein. This period was 1.1 million years for proteins known as fibrinopeptides, 5.8 for hemoglobins, 20 for cytochromes *c*, and about 500 million years for histone 4. Remarkably, the nucleotides in the genes coding for these proteins all changed at about the same rate. Although Dickerson (1971) did not differentiate between one-base and two-base changes in amino acid replacements, it is clear that he perceived this regularity of rate, because he wrote: "What is observed is the rate of mutation of the nucleic acids, *modified by the elimination of nonfunctional molecules*" (Dickerson, 1971; emphasis added). Dickerson's diagram is shown in simplified form in Figure 2.3.

"Elimination of nonfunctional molecules" suggests that huge numbers of mutations in genes for hemoglobin, cytochrome *c*, and especially for histones, are deleterious or lethal. This inference was substantiated when the DNA molecules for these genes were sequenced, and silent substitutions were found to predominate, implying that a large number of nonsilent changes had been eliminated by natural selection.

Globins

Hemoglobins have been used more than any other protein in studies of molecular evolution. Globin proteins occur in blood and transport oxygen in vertebrates from the lungs

FIGURE 2.3

Rate of evolution of three different proteins, as measured by unit evolutionary periods (millions of years/1% change in amino acid sequence). (Modified from Dickerson, 1971.)

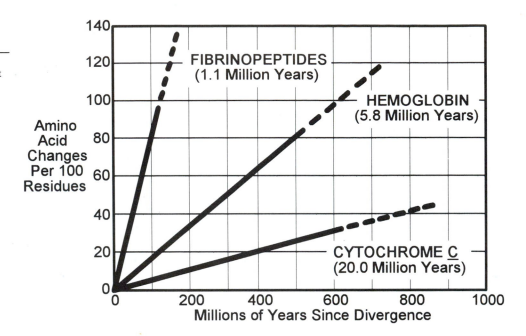

or gills to the other tissues. All globin genes are derived from a single ancestral gene. Red blood cells contain a globin **tetramer**, a four-component molecular grouping that consists of two identical alpha (α) and two beta (β) hemoglobins. The genes for human α globins are carried on chromosome number 16, and those for β on number 11. The genes occur in clusters of six or seven (Figure 2.4), and the β cluster includes two β-like gamma (γ) genes that are used in fetal hemoglobin, in which the tetramer consists of two α and two γ hemoglobins. The γ chains are replaced by β chains during infancy (Figure 2.5). Another globin, called myoglobin, is a single-component **monomer** that is present in muscle; its gene was derived from the α globin line of descent about 750 million years ago.

Duplication of the ancestral α globin gene into genes for α and β took place about 600 million years ago. This gene-duplicating event—a rather common occurrence in molecular evolution (see Chapter 4)—led to the formation of the tetramers described above, and increased the efficiency both of the uptake and the release of oxygen. This was a beneficial evolutionary event that enhanced the ability for "fight or flight." From then on, the α and β chains evolved separately, but they retained the capacity to form loosely connected α-β pairs in all vertebrate animals.

Silent substitutions take place at silent sites in protein-coding genes. As shown in Figure 2.2, there are four types of silent sites (Jukes and Bhushan, 1986):

1. Third-base positions of 61 codons (all 64 triplets except for the three stop codons), minus G in third positions of methionine and tryptophan codons

2. A in first-base positions of A-G-R arginine codons (in which R represents either A or G), that is, the A-G-A and A-G-G codons for arginine

3. C in first-base positions of C-G-R arginine and C-T-R leucine codons

4. U in first-base positions of U-U-R leucine codons

In total, there are 67 potential sites for silent substitutions.

In comparisons of two homologous genes, silent substitutions are counted as illustrated in the hypothetical example shown in Box 2.1. In such comparisons, any nucleotide substitution that produces an amino acid change is defined as a replacement.

Table 2.1 shows that the number of silent changes per 100 codons increases with the

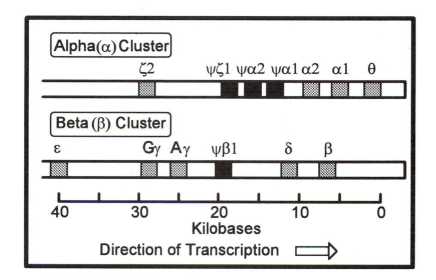

FIGURE 2.4

Alpha-like and beta-like globin gene families organized into clusters. Each cluster includes functional genes (shaded boxes) and pseudogenes (black boxes). The organization of the clusters in higher primates is conserved (unchanged during evolution); therefore, the clusters in humans, baboon, and orangutan are virtually indistinguishable. There are two human γ genes, Aγ and Gγ; their products are identical, except that the Aγ protein has alanine at position 136 and the Gγ protein has glycine. All active genes are transcribed (that is, their message is transferred to mRNA) from left to right. (Modified from Dickerson and Geis, 1983.)

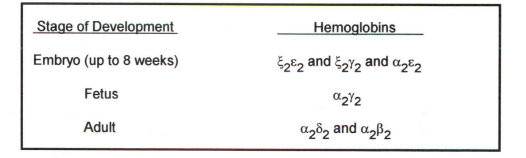

FIGURE 2.5

Changes in human hemoglobins during development. The first α-like chain to be expressed is ξ, but it is soon replaced by α itself. In the β-pathway, ϵ and γ are expressed first, with δ and β replacing them later. In adults, the $\alpha_2\beta_2$ form provides 97% of the hemoglobin, $\alpha_2\delta_2$ provides approximately 2%, and persistence of the fetal form, $\alpha_2\gamma_2$, provides approximately 1%. (Modified from Dickerson and Geis, 1983.)

BOX 2.1

Hypothetical example illustrating the way that silent substitutions and replacements of amino acids are counted in a comparison of two homologous genes.

Arrows show changes that occur in the nucleotide nitrogenous bases as one codon mutates into another.

HOMOLOGOUS GENE 1

mRNA codons:	A-G-A	C-G-G	C-U-A	A-U-U	G-U-C	U-C-A
Amino acids:	↓ Arg	↓ Arg	↓ Leu	Ile ↓	Val ↓	↓ Ser

Intermediate Changes   ↓   ↓   ↓   ↓   ↓   ↓

mRNA codons:	C-G-A	↓	U-U-A	↓	G-U-U	↓
Amino acids:	Arg ↓	↓	Leu ↓	↓	↓ Val	↓

HOMOLOGOUS GENE 2   ↓   ↓   ↓   ↓   ↓   ↓

mRNA codons:	C-G-G	A-G-G	U-U-G	A-U-C	A-U-U	G-C-A
Amino acids:	Arg	Arg	Leu	Ile	Ile	Ala

SILENT CHANGES:	2	1	2	1	1	0
REPLACEMENTS:	0	0	0	0	1	1

amount of evolutionary distance that separates the two species—and, therefore, with the amount of time that has elapsed since the divergence of the compared lineages—which is to be expected from the corresponding increase in the numbers of mutations that will have occurred as the time since divergence increases. The number of replacement changes, in contrast, is unrelated to time, because it depends on whether the changes are acceptable to protein function. Note that replacements in hemoglobin are relatively common, in contrast to replacements in cytochromes *c* and, especially, histone 3. Hemoglobin can accept

TABLE 2.1

Comparison of silent changes and amino acid replacement nucleotide substitutions in homologous proteins from various organisms.

		Changes/100 Codons/100 Million Yr	
Homologous Protein	**Organisms Compared**	**Silent Changes (no.)**	**Replacements (no.)**
α hemoglobin	rabbit:mouse	37	22
β hemoglobin	rabbit:mouse	34	25
Cytochrome *c*	rodent:chicken	43	8
Histone 3	sea urchin:trout	50	0.8

many amino acid changes; in fact, in the many hemoglobins that have been analyzed, there are only two amino acid sites that do not vary from one species to another.

Figure 2.6 shows, diagrammatically, the **barrier effect** in molecular evolution, in which nucleotide substitutions "hit the barrier" of natural selection. The silent changes are neutral, and because natural selection therefore does not select against them, they pass through the barrier. Replacement changes are screened by natural selection; whether they pass through the barrier depends on the nature and extent to which they alter the properties of the protein and are thereby either advantageous (rare), more or less neutral, or deleterious or lethal (more common) to the organism. The rate of mutation in these coding regions of the DNA is the same as that resulting in silent nucleotide changes.

For hemoglobin, the constraint imposed by natural selection is fairly weak. About 22% of the replacement changes are accepted per 100 million years, as compared with 37% of silent changes (Table 2.1). Cytochrome c is more tightly constrained, indicating that a greater proportion of amino acid replacements are deleterious or lethal, and are therefore selected against; in this protein, only eight replacement changes per 100 codons are accepted per 100 million years. Histones accept practically no amino acid replacements (less than 1% per 100 million years) because of their rigidly defined chromosome-associated function.

The number of base differences per codon in comparisons of homologous proteins can be explained in great detail with today's knowledge. For example, Collins and Jukes (1994) recently compared 337 pairs of homologous genes in humans and rodents, more than 86% of which were identical at the amino acid level (Table 2.2). The comparison included more than 159,000 codon pairs, having a mean base difference per codon of 0.165. Pairs of codons for the eight "family-box" amino acids (discussed later) were studied in some detail: about 34% have been substituted at the third-base (silent) position. This value is an average for all 337 pairs of homologous genes, and values for particular gene-pairs encompass a consid-erable range. Collins and Jukes (1994) also found that amino acid replacements in the family-box pairs were only about 8% as compared with more than 17% for non-family-box pairs, and that there is a lower percentage of two-base changes (15%) in family-box replacements as compared with non-family-box replacements (22%).

In general, these observations favor the explanation of a steady incidence of nucleotide substitutions distributed approximately at random and arising from errors in DNA replication. This accounts for the gradual increase in two-base changes as genes diverge and for the spread of substitutions throughout amino acid codes.

WHAT ARE NEUTRAL CHANGES?

Junk DNA

In 1969 King calculated that probably not much more than 1% of mammalian DNA coded for proteins, that is, nearly 99% of the DNA in mammalian chromosomes is "non-coding" (King and Jukes, 1969). This was a new idea. A year earlier, Kimura (1968) had assumed that the *entire* genome of mammals had the same evolutionary rate as that of the average of three proteins: hemoglobin, cytochrome c, and triosphosphate dehydrogenase. The suggestion that about 99% of mammalian DNA was non-coding (**junk DNA**) was both provocative and predictive. The current estimate (Nowak, 1994) is that 97% of the DNA in humans *"does not code for proteins or for RNA with clear functions"* (such as those performed by mRNA, transfer RNA, and ribosomal RNA). Nowak (1994) listed the following nine types of "junk" nucleic acids.

FIGURE 2.6

Diagrams illustrating how mistakes in DNA replication produce nucleotide substitutions that can be accepted or rejected in evolution. Comparisons of homologous hemoglobin molecules in rabbit and mouse and homologous cytochrome *c* molecules in rabbit and chicken show that some substitutions pass through the barrier of natural selection and are accepted either as amino acid replacements or as silent changes. In contrast, comparison of homologous histone 3 molecules in sea urchin and trout shows that only silent changes pass through the barrier and that virtually no amino acids are replaced.

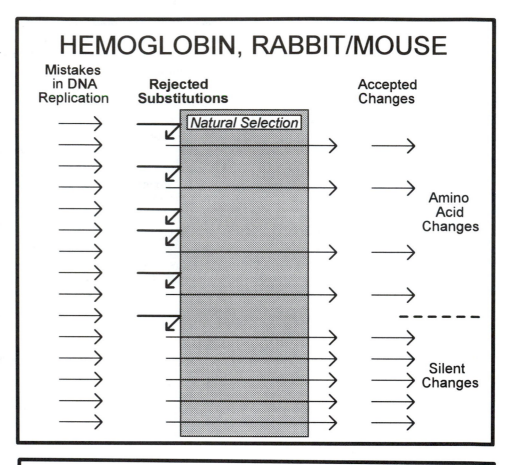

HEMOGLOBIN, RABBIT/MOUSE

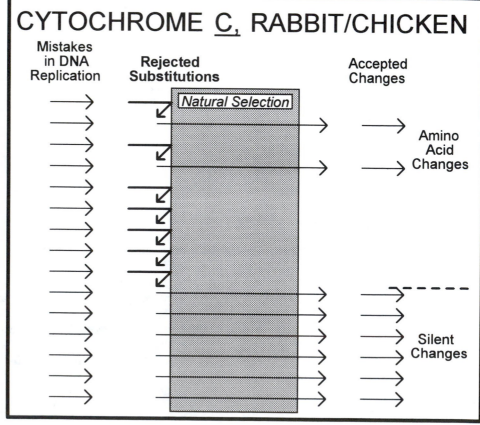

CYTOCHROME C, RABBIT/CHICKEN

FIGURE 2.6

Continued

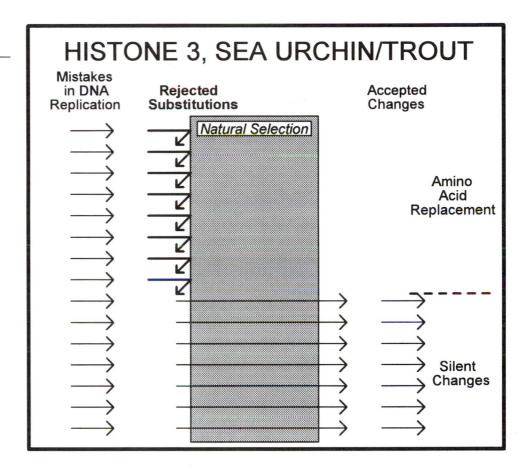

1. *Introns*: Long stretches of non-coding DNA within genes. These segments interrupt the protein coding regions and must be excised before the gene message is encoded into mRNA.

2. *Satellites*: Short DNA sequences that are repeated hundreds of times, mainly at the ends and centers of chromosomes.

3. *Minisatellites*: Sequences that are shorter than satellites, repetitive, and dispersed throughout chromosomes.

TABLE 2.2

Codon changes in 337 pairs of homologous human/rodent genes.

Total no. of codons in alignments, pairs	159,063
Amino acid replacement	21,570
Single-base change	17,117
Two-base change	4,290
Three-base change	163
Percentage of replacement	13.6
Mean base difference/codon	0.165

Source: Data from Collins and Jukes (1994).

4. *Microsatellites*: Sequences that are similar to, but shorter than, minisatellites.

5. *3'-Untranslated regions*: Regions at one end of mRNA molecules that do not appear to function in protein synthesis.

6. *Heterogeneous nuclear RNA*: RNA occurring in the cell nucleus that has no known function.

7. *Short interspersed elements*: These elements can cause disease if interspersed within a gene (e.g., the "Alu" sequence discussed in Chapter 1).

8. *Long interspersed elements*: Elements as much as 7,000 base pairs in length that are capable of causing disease.

9. *Pseudogenes*: Defective copies of normal genes that because of their defects are unable to function. Sometimes described as "decaying, rusted hulks," they evolve as if they were silent or neutral. Their evolutionary rate is about 5×10^{-9} per site per year, a value that probably approximates the maximum rate of neutral evolution.

Because this junk DNA is non-coding, it does not produce protein products, and there are therefore no products on which natural selection can operate. The junk DNA can mutate, but this has no effect on the organism. This is not true of DNA changes that result in amino acid replacements in proteins, which are constrained by natural selection. Thus, changes in amino acid sequences occur much more slowly than changes in total DNA.

Neutral Changes Are the Source of the Molecular Evolutionary Clock

Different hemoglobin molecules occur in various vertebrates (Table 2.3). These have amino acid differences ranging more than 60% in the comparison of bony vertebrates with sharks (which lack true bones). If these differences are the result of adaptation and natural selection, they should be dictated by the living requirements of the organism.

Max Perutz stated (1983) that the complicated (tertiary and quaternary) structures of deoxy- and oxyhemoglobin have remained almost unchanged during vertebrate evolution, and that most of the amino acid replacements that occurred during the evolution of vertebrate hemoglobins are functionally neutral.

Despite the characteristic complexity of the structure of the hemoglobin molecule (including the presence of a very distinctive, internal, essentially water-free hydrophobic cavity), it can accommodate large numbers of amino acid replacements. Perutz (1983) pointed out that, in hemoglobin, adaptations leading to responses to new chemical stimuli have evolved by the occurrence of only a few (one to five) amino acid replacements in key positions in the molecule. Therefore, more or less the same small number (five to ten) of sequence differences should occur between hemoglobins in any pair of vertebrates, such as between humans and frogs or between sharks and snakes, and the sequence differences should be expressed in individual species and scattered irregularly among members of the various vertebrate lineages. But this is not the case. Indeed, there has been a steady and increasing divergence of the sequences over the course of vertebrate evolution (Table 2.3), explainable because neutral changes accumulate in hemoglobin at a fairly uniform rate. Thus, the neutral process of amino acid substitution gives rise to a **molecular evolutionary clock**—the amount of sequence difference is approximately equivalent to the amount of time that has elapsed since the two lineages diverged—a clocklike behavior (see Chapter 3) that would not exist without neutral changes.

The Amino Acid Code

At the molecular level, mutations in the DNA are related to the amino acid code (Figure 2.2) that is carried by the mRNA. A point mutation in the DNA is typically reflected by a change in a single nucleotide in a codon for an amino acid. The codons are units of three nucleotides. There are 64 such units, 61 for amino acids and the other 3 encoding "stop" messages that bring about the termination of the synthesis of a protein molecule by interacting with a **release factor**. This interaction separates the protein molecule from the ribosomal site of protein manufacture.

Any change in the second position of a codon, and nearly all changes in the first position, bring about the replacement of an amino acid. For example, if the codon U-U-U is changed to U-C-U, the amino acid inserted into the protein is changed from phenylalanine to serine. However, none of the U to C interchanges in the third-base position affects the protein product and, as discussed earlier, they are therefore called silent changes. The code shown in Figure 2.2 contains 20 amino acids, but the codons are divided neither equally nor randomly among them. There are 16 groups (boxes) in the table, and 8 of them are occupied by one individual amino acid. These 8 boxes are called "**family boxes**." Some years ago, Jukes (1966) suggested that in an earlier-evolving, more primitive code there were 15 family boxes for 15 amino acids, and a single box for stop codons. Recently, Lehman and Jukes have proposed that earlier-evolved codes may have contained a large number of stop codons (of which, today, there are only three), and that the code evolved by assignment of these codons to new amino acids (Lehman and Jukes, 1988). If this is correct, reassignment of these codons would have led to an increase in the number of amino acids used in protein synthesis.

TABLE 2.3

The molecular evolutionary clock in alpha and beta hemoglobins.

Organisms Compared	Hemoglobin Amino Acid Changes/100 Codons			Approximate Time since Divergence (Million Yr)
	$\alpha*$	$\beta*$	$\beta\dagger$	
Human:Monkey			5	20
Placental mammals (between species)	16	17		100
Human:Cow			18	
Kangaroo:Placental mammals	22	27		160
Platypus:Placental mammals			25	
Chicken:Mammals	30	32	32	215
Snake:Warm-blooded vertebrates	39			290
Amphibians:Terrestrial vertebrates	47	49		380
Carp:Tetrapod vertebrates	49	50		400
Shark:Bony vertebrates	58	64	65	545

*Data from vertebrate animals reported by Jukes (1980).

†Data from vertebrate animals reported by Dickerson and Geis (1983).

The amino acid sequence of a protein can be read from the nucleotide sequence in a gene by means of the code. This is a very important procedure used in studies of molecular evolution. However, because of the redundancy of the code, the reverse is not true—almost all amino acids are coded for by several different codons, and if only the amino acid sequence is known, it is impossible to deduce the exact nucleotide sequence in the DNA. Fortunately, it is easier to sequence DNA than to sequence proteins. The usual procedure is, first, to sequence the DNA; second, to look for **open reading frames** (long stretches of DNA sequences without stop codons); and third, to search for the protein products. By the rules of chance, on average, every twenty-first triplet in a non-coding DNA sequence should carry the message for a stop codon. If this is found, it is not an open reading frame, instead, it is a region of DNA that does *not* contain genes for proteins. As discussed earlier, most DNA is like this, the "junk DNA."

The genetic code is related to practically all studies of molecular evolution. Once it became known, comparing protein sequences and drawing conclusions as to what had happened during the evolution of their genes became possible.

Progress of the Neutral Theory

In evolution, neutral amino acid replacements that do not change the function of proteins have to spread through populations. Studies of population genetics show that a change that helps an organism to adapt and survive may enable that organism to displace others. If two species differ in their cytochromes c, some way for differences to spread through their species populations must occur. In 1964, biologist George Gaylord Simpson stated that it was highly improbable that proteins, fully determined by genes, should have non-functional parts, and in 1967, another scientist asserted that each amino acid of a protein molecule must have a unique survival value to the phenotype of an organism.

An obvious location for neutral changes was in DNA, not in proteins. The gene for a sequence of amino acids can change virtually all its third-base codon positions without changing the amino acid sequence (Jukes, 1965). During that period, most classic evolutionists were not familiar with the genetic code. King and Jukes (1969) claimed, however, that because of the redundancy (**degeneracy**) of the genetic code, some DNA base-pair changes in structural genes are without effect on protein structure. There are 61 amino-acid-specifying codons and because each of the three base pairs in these codons can mutate in any of three different ways (to each of three different base-containing nucleotides), each codon can mutate in any of nine ways by a single substitution. Of the 549 possible single-base substitutions, 134 (one-fourth) are substitutions to synonymous codons (Jukes, 1965; King and Jukes, 1969). These are heritable changes in the genetic material, and they are therefore true mutations. Thus, synonymous mutations seem to be truly neutral with respect to natural selection.

At the end of the 1960s, the **Principle of Neutral Substitutions** was developed. Jukes and King compared sequence changes in proteins and in DNA by conducting **DNA-DNA hybridization** experiments. Heating DNA gently at about 65°C causes its two strands to separate (melt). After slowly cooling, the strands reunite, a natural result of the ability of one strand to link to another by hydrogen bonding (Figure 2.1). For example, if one strand of rat DNA is mixed, before it cools, with heated DNA from a mouse, some hybrid rat-mouse strands will form, partly joined together by bonds between some, but not all, pairs of bases. This is **hybrid DNA**, and it has a lower melting point—that is, it is easier to separate into two strands by heating—than either unhybridized rat or mouse DNA. About 15% of the nucleotides in this hybrid are not base-paired with bases in the other strand. Some of these unpaired bases are in the first two base codon positions, but a substantial number of them are unpaired in third-base codon positions. If the third-base codon positions are the same in homologous portions of the DNA, the hybrid will be rather tightly bonded and have a high melting point, but if these bases have changed, the two strands will be loosely bonded and the hybrid will have a lower melting point. The strength of the expected hybridization can be calculated from the difference in amino acids at various positions in homologous proteins, such as hemoglobins from rats and mice, or from nucleotide differences expected on the basis of the time elapsed since the evolutionary divergence of the organisms being compared. Experiments of this type enabled Jukes and King to conclude that the third position of the codons had changed more rapidly than either of the first two positions.

The idea of nonfunctional amino acid replacements in proteins ran counter to traditional thinking. Classic evolutionary theory did not include the idea of silent substitutions in third codon positions. The classic viewpoint was that because such changes would not affect the organismal phenotype, they would not occur. Simpson (1964), a leading proponent of this view, had stated that

> The consensus is that completely neutral genes or alleles must be very rare if they exist at all. For an evolutionary biologist, it seems highly unprobable that proteins, supposedly fully determined by genes, should have nonfunctional parts . . . or that molecules should change in a regular but nonadaptive way.

Jukes and King were of the opinion that this traditional "consensus" view needed to be examined in light of newly discovered molecular facts.

Simpson concluded the passage quoted above by stating that natural selection

> is the composer of the genetic message, and DNA, RNA, enzymes and other molecules in the system are successively its messengers.

King and Jukes (1969) had a different view, namely that

> evolutionary change is not imposed on DNA from without, it arises from within. One thing the editor does not do is to remove changes which it is unable to perceive.

The essence of their theory was that **the driving force of evolution is DNA, through changes occurring in the DNA molecule**.

The history of the development of the neutral theory, including Kimura's (1968) contributions, has been summarized recently (Jukes, 1991). According to this account, the article by King and Jukes (1969) was initially rejected by both of the reviewers (anonymous scientific colleagues) to whom it had been assigned. Ironically, one reviewer is reported to have viewed the theory as lacking credibility, whereas the other suggested that it was too obvious to merit publication. King and Jukes appealed the decision to the journal editors (an appeal that today would not be permitted by the editorial policy of *Science*), and the manuscript was ultimately accepted for publication. After publication, the title of the article, "Non-Darwinian Evolution" (coined by King), was much criticized because it implied that important aspects of biological evolution could be other than Darwinian.

Multiple Substitutions

As evolution progresses, the laws of chance hold that some nucleotide sites will be mutated more than once, such as from A to G, then from G to C; or from A to G, then from G back to A. The second of these mutations is called a **revertant change**; that is, after a change from A to G, the G reverts again to A. An equation for calculating the predictability of such multistep changes was derived by Charles Cantor (Jukes and Cantor, 1969). The mean number of base differences at a single position in the mRNA (μ) is related to the observed fraction of residues with single-base differences (ρ) by the expression:

BOX 2.2

Revertant and multistep changes in homologous sequences of human alpha and beta hemoglobin genes.

> A 30-nucleotide portion of the homologous sequences of the human α and β hemoglobin genes
>
> α gene: A-C-*C*-*A*-*A*-C-G-T-*C*-A-*A*-*G*-G-C-C-*G*-*C*-*C*-T-G-G-G-G-*T*-A-A-G-G-T-*T*
> β gene: *T*-C-*T*-*G*-*C*-C-G-T-*T*-A-*C*-*T*-G-C-C-*C*-*T*-*G*-T-G-G-G-G-*G*-A-A-G-G-T-*G*
>
> Of these 30 sets of nucleotides, 12 (shown in ***bold italics***) differ between the two sequences; thus, 12/30 (40%) of the sites show nucleotide substitution. However, the mean number of substitutions that has actually occurred must be greater than 12, because of revertants and multistep changes. The equation, $\mu = 3/4 \ln [3/(3-4\rho)]$–in which "$\mu$" is the mean number of base differences at a single position on the mRNA, and "ρ" is the observed fraction of residues with single-base differences–corrects for these, indicating that the probable total number of substitutions is 17 (57%), higher than the 12 (40%) shown in the alignment.

$$\mu = 3/4 \ln [3/(3-4\rho)]$$

The equation assumes that all single-base changes (that is, *all* nucleotide substitutions) are equally probable, and that the frequencies of occurrence for all four bases in DNA are the same. If these conditions are met, there is one chance in four that any two nucleotides will be identical when two unrelated sequences of A, C, G, and T are compared. Cantor's formula came into wide use when rapid DNA and RNA sequencing became available in the 1970s. Since then, molecular biologists have become increasingly interested in comparing sequences of homologous genes to study evolution. The example in Box 2.2 shows the use of this technique.

■

SUMMARY

The neutral theory of molecular evolution postulates that most accepted mutations have little or no selective effect on the survival of an organism. In addition to accepted mutations, there are many deleterious mutations that are not accepted. The evidence for the neutral theory is that silent mutations, such as nucleotide substitutions in the synonymous positions of amino acid codons, occur commonly. Such substitutions are increased by directional mutation pressure.

Hemoglobin molecules have undergone many amino acid replacements during their evolution, but have retained their structure and function. Only two amino acid residues in hemoglobin do not vary from one species to another. Because hemoglobin has retained its structure and function, despite this striking variability, the amino acid replacements must be neutral or near-neutral. The steady, time-related rate at which such changes have been accepted into functioning molecules has given rise to the concept of the molecular evolutionary clock.

Silent nucleotide substitutions are accepted more rapidly than are amino acid replacements, as shown by the evolution of various types of proteins. One of the most extreme examples is that of histones, which have changed very little in their amino acid composition over evolutionary history, although silent substitutions in their genes are accepted at about the same rate as in other proteins, such as hemoglobins. In pseudogenes, the most rapid rate at which silent nucleotide substitutions are accepted is about 5×10^{-9} per site per year, equivalent to one change per site per 200 million years. At this rate, about 70 silent substitutions would be expected to show in a comparison of rabbit and mouse α hemoglobin genes; the actual number found is 53 per 141 sites, close to the predicted range.

Mayr (1978) defined evolution as "change in the diversity and adaptation of populations of organisms." Neutral changes affect diversity at the molecular level, but they do not affect adaptation. Adaptation rejects deleterious (including lethal) changes and accepts advantageous changes. Neutral changes, which are neither deleterious nor advantageous, enter populations by genetic drift, a well-defined and much studied feature of population genetics. The driving forces in evolution are errors in DNA replication, gene duplication, and other molecular changes in DNA. Among these, errors in replication are the most frequent.

Acknowledgments

I wish to thank my colleagues, especially Charles Cantor, Jack King (who died of leukemia in 1983), Vikas Bhushan, Niles Lehman, and Dave Collins.

REFERENCES

Collins, D., and Jukes, T.H. 1994. Rates of transition and transversion in coding sequences since the human-rodent divergence. *Genomics 20:* 386–396.

Connor, W., States, J.C., Mezquita, J., and Dixon, G.H. 1984. Organization and nucleotide sequence of rainbow trout histone H2A and H3 genes. *J. Mol. Evol. 20:* 236–250.

Cox, E.C., and Yanofsky, C. 1967. Altered base ratios in the DNA of an *Escherichia coli* mutator strain. *Proc. Natl. Acad. Sci. USA 58:* 1895–1902.

Dickerson, R.E. 1971. The structure of cytochrome c and the rates of molecular evolution. *J. Mol. Evol. 1:* 26–45.

Dickerson, R.E., and Geis, I. 1983. *Hemoglobin* (Menlo Park, CA: Benjamin-Cummings).

Freese, E. 1962. On the evolution of base composition of DNA. *J. Theoret. Biol. 3:* 82–101.

Horowitz, N.H. 1994. The molecular vision of life. *Biophys. J. 66:* 929–930.

Jukes, T.H. 1963. Some recent advances in studies of transcription of the genetic message. In: H. Coghill (Ed.), *Advances in Biological and Medical Physics 9* (New York: Academic Press), pp. 1–41.

Jukes, T.H. 1965. The genetic code II. *Am. Scientist 53:* 477–487.

Jukes, T.H. 1966. *Molecules and Evolution* (New York: Columbia Univ. Press).

Jukes, T.H. 1980. Silent nucleotide substitutions and the molecular evolutionary clock. *Science 210:* 973–978.

Jukes, T.H. 1991. Early development of the neutral theory. *Persp. Biol. Med. 34:* 473–485.

Jukes, T.H., and Bhushan, V. 1986. Silent nucleotide substitutions and G + C content of some mitochondrial and bacterial genes. *J. Mol. Evol. 24:* 39–44.

Jukes, T.H., and Cantor, C.R. 1969. Evolution of protein molecules. In: H.N. Munro (Ed.), *Mammalian Protein Metabolism. 3* (New York: Academic Press), pp. 21–132.

Kimura, M. 1968. Evolutionary rate at the molecular level. *Nature 217:* 624–626.

King, J.L., and Jukes, T.H. 1969. Non-Darwinian evolution. *Science 164:* 788–798.

Lehman, N., and Jukes, T.H. 1988. Genetic code development by stop codon takeover. *J. Theoret. Biol. 135:* 204–214.

Mayr, E. 1978. Evolution. *Scientific American 239:* 47–55.

Mayr, E. 1993. The resistance to Darwinism and the misconceptions on which it was based. In: J.H.

Campbell and J.W. Schopf (Eds.), *Creative Evolution?!* (Boston, MA: Jones & Bartlett), pp. 35–46.

Nowak, R. 1994. Mining treasures from junk DNA. *Science 263:* 608–610.

Perutz, M. 1983. Species adaptation in a protein molecule. *Mol. Biol. Evol. 1:* 1–28.

Schaffner, W., Kunz, G., Daetwyler, H., Telford, J., Smith, H.O., and Birnstiel, M.L. 1978. Genes and spacers of cloned sea urchin histone DNA analyzed by sequencing. *Cell 14:* 655–671.

Simpson, G.G. 1964. Organisms and molecules in evolution. *Science 146:* 1535–1538.

Sueoka, N. 1961. Correlation between base composition of deoxyribonucleic acid and amino acid composition of protein. *Proc. Natl. Acad. Sci. USA 47:* 1141–1149.

Sueoka, N. 1962. On the genetic basis of variation and heterogeneity of DNA base composition. *Proc. Natl. Acad. Sci. USA 48:* 582–592.

Sures, I., Lowry, J. and Kedes, L. 1978. The DNA sequence of sea urchin (*S. purpuratus*) H2A, H2B and H3 histone coding and spacer regions. *Cell 15:* 1033–1044.

Watson, J.D., and Crick, F.H.C. 1953. Molecular structure of nucleic acids. *Nature 171:* 738–740.

Winkfein, R.J., Connor, W., Mezquita, J., and Dixon, G.H. 1985. Histone H4 and H2B genes in rainbow trout (*Salmo gairdnerii*). *J. Mol. Evol. 22:* 1–9.

CHAPTER 3

GENES, SEQUENCES, AND CLOCKS: MOLECULAR CLUES TO THE HISTORY OF LIFE

■

Bruce Runnegar*

■

INTRODUCTION

Informative Life

Genes are the software of life. The data files and programs of the cellular operating system are written in a low-level language and stored on tapes made of DNA. This machine language has a four-letter alphabet of DNA **nucleotides** (A, C, G, T), a limited vocabulary of short words (amino acid **codons** such as CAT or GAG and gene switches such as TATAAA), a few common dialects, and simple rules for the construction of sentences. Nevertheless, it is rich enough to build an elephant from a single cell. It is this **information content** that is most characteristic of living things; all other criteria for the presence of life are insignificant beside it.

Life's incessant accumulation of information goes against the grain of the universe. Whereas most natural processes dissipate energy and destroy order, life continually creates order (using up solar energy in the process). However, the spontaneous emergence of complexity, and hence order, is now thought to be a property of many natural processes (Waldrop, 1992). Such systems operate at the edge of chaos in that they lie, mathematically, between the readily predictable (ocean tides, phases of the Moon, compound interest) and unpredictable chaos (whitewater turbulence, next month's weather, stock market peaks and troughs).

The information content of Earth life has grown steadily over the past 4 billion years. It may now be as large as 10^{15} **quits** (quaternary digits, comparable to binary digits or bits, 0 and 1, used by computers but based on the four nucleotide states rather than on two electronic ones). This figure is obtained by multiplying the number of existing species (e.g., 10^7) by the number of meaningful nucleotides in an average **genome** (e.g., 10^8). The number is surprisingly small (one million gigaquits) compared with the number of bits (or bytes) that are currently stored on the world's computer hard drives. For example, a

*Department of Earth and Space Sciences, Molecular Biology Institute, and Institute of Geophysics and Planetary Physics, University of California, Los Angeles, California 90095.

data silo installed recently at the Australian National University's Computer Services Centre has a storage capacity of about 240 terabytes (2.4×10^{14} bits). The biosphere's one million gigaquits is equivalent to 250 terabytes.

If we could tap this store of information we should be able to unravel the history of life. Although new information is always being created, much is inherited from the past. The important information is that retained and modified in different lines of descent. Technically, the best information is both *shared and modified*. In the words of **cladists** (scientists who study relationships among organisms; see also Chapter 1), to understand evolution the search is for **synapomorphies** (shared derived characters) not **symplesiomorphies** (shared primitive characters) nor **autapomorphies** (unique characters found in only one line).

The difficulty with using the cellular machine code for historical purposes is that the number of available character states is small (A, C, G, T in DNA, and A, G, C, U in RNA). Shared states resulting from convergent evolution (**homoplasy**) are difficult to distinguish from those derived by inheritance (**homology**). This problem is particularly serious for the exploration of deep time (e.g., more than 1 billion years ago) because the information may have been modified many times. However, there are ways to estimate and correct for the amount of overprinting that has occurred within genes.

FIGURE 3.1

The universal genetic code (dialects such as UGA instead of UGG for tryptophan are found in a few organisms and some organelles). The nucleotide bases are either a pyrimidine (C or U, a hexagon of carbon and nitrogen atoms) or a purine (A or G, a hexagon and a pentagon of carbon and nitrogen atoms). The DNA of a gene is first copied (transcribed) into a messenger RNA (mRNA), then the mRNA is translated into protein using the genetic code. Each amino acid is coded by one codon of three bases (at first, second, and third positions). The amino acids are commonly represented by one letter or three letter abbreviations as shown on the right (A or Ala for Alanine, and so forth).

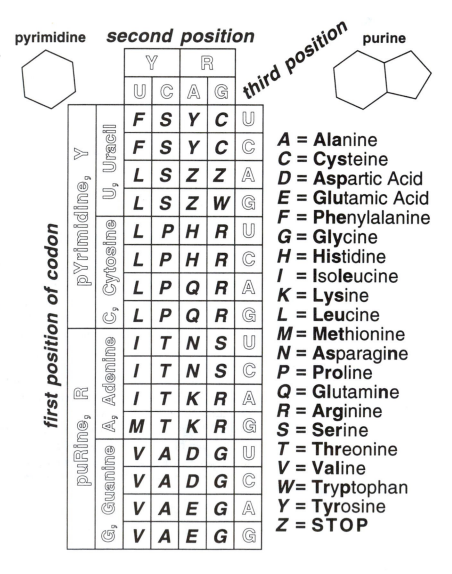

A = **Ala**nine
C = **Cys**teine
D = **Asp**artic Acid
E = **Glu**tamic Acid
F = **Phe**nylalanine
G = **Gly**cine
H = **His**tidine
I = **Ile**ucine
K = **Lys**ine
L = **Leu**cine
M = **Met**hionine
N = **Asp**aragine
P = **Pro**line
Q = **Glu**tamine
R = **Arg**inine
S = **Ser**ine
T = **Thr**eonine
V = **Val**ine
W = **Trp**tophan
Y = **Tyr**osine
Z = **STOP**

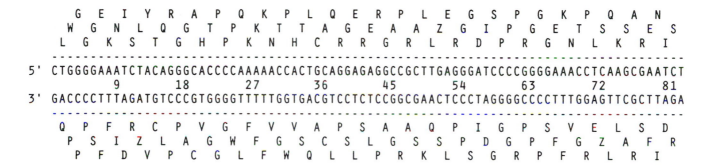

```
  1 CTGGGGAAAT CTACAGGGCA CCCCAAAAAC CACTGCAGGA GAGGCCGCTT GAGGGATCCC

 61 CGGGGAAACC TCAAGCGAAT CTGGGAAGGG AGCGTACCTG GGTCGATCGT GCGCGTTGGA

121 GGAGACTCCT TCGTAGCTTC GACGCCCGGC CGCCCCTCCT CGACCGCTTG GGAGACTACC

181 CGGTGGATAC AACTCACGCG GCTCTTACCT GTTGTTAGTA AAAAAAGGTG TCCCTTTGTA

241 GCCCCT
```

```
     G   E   I   Y   R   A   P   Q   K   P   L   Q   E   R   P   L   E   G   S   P   G   K   P   Q   A   N
     W   G   N   L   Q   G   T   P   K   T   T   A   G   E   A   A   Z   G   I   P   G   E   T   S   S   E   S
     L   G   K   S   T   G   H   P   K   N   H   C   R   R   G   R   L   R   D   P   R   G   N   L   K   R   I
     -----------------------------------------------------------------------------------------------------
  5' CTGGGGAAATCTACAGGGCACCCCAAAAACCACTGCAGGAGAGGCCGCTTGAGGGATCCCCGGGGAAACCTCAAGCGAATCT
         9        18        27        36        45        54        63        72        81
  3' GACCCCTTTAGATGTCCCGTGGGGTTTTTGGTGACGTCCTCTCCGGCGAACTCCCTAGGGGCCCCTTTGGAGTTCGCTTAGA
     -----------------------------------------------------------------------------------------------------
     Q   P   F   R   C   P   V   G   F   V   V   A   P   S   A   A   Q   P   I   G   P   S   V   E   L   S   D
     P   S   I   Z   L   A   G   W   F   G   S   C   S   L   G   S   S   P   D   G   P   F   G   Z   A   F   R
     P   F   D   V   P   C   G   L   F   W   Q   L   L   P   R   K   L   S   G   R   P   F   R   L   R   I
```

FIGURE 3.2

The cadang-cadang coconut viroid genome. This genome is a loop of 246 DNA nucleotides, each in one of four possible states: A, G, C, T (top panel). It therefore represents 246 quaternary digits (quits) of information and, theoretically, could code for 492 amino acids (82 in each of six reading frames); those corresponding to the first 82 nucleotides are shown using the one-letter abbreviations for amino acids above the second position of the relevant codons (bottom panel); Z = stop, a codon that signals the molecular machinery that this is the end of the protein.

Fortunately, there is also a second, higher-level language that is used by all cells. It is obtained by translating parts of the DNA machine code into protein applications (Figure 3.1). The advantage of using this amino acid-based language for historical purposes is that it has an alphabet of 20 letters instead of four (A, C, D, E, F, G, H, I, K, L, M, N, P, Q, R, S, T, V, W, Y); consequently, homoplasy is rarer and easier to detect. Finally, because the protein alphabet consists of 20 kinds of amino acids, the words are larger building blocks, known as **exons**, **motifs**, and **domains**. These components are the building blocks of life.

Evolving Genomes

Genes are the regions of the genome that are copied (**transcribed**) into RNA. They may be separated by long intergenic regions, be placed end to end, or even be superimposed in overlapping reading frames (Figure 3.2). Thus, the fraction of genes in a genome varies greatly, from a few percent in the exceptionally large genomes of ferns and lungfish (10 to 100 billion nucleotides), through about 80% in the pufferfish (unusually small for a vertebrate) to more than 100% (111.2%) in the bacterial virus ϕX174 in which genes overlap in the six different reading frames that are available in any DNA double helix (see Figure 3.2). Most human genes and those of other complex organisms (higher eukaryotes) are interrupted by **introns**. These are lengths of noncoding DNA that are transcribed and then excised from the immature mRNA by a process called **splicing**; after splicing has occurred, the originally separate coding regions of the gene (**exons**) are joined end to end.

When introns were discovered, they were explained in two main ways. One suggestion, made independently by James Darnell and W. Ford Doolittle, was that introns were present in all early genes but were shed in lines leading to modern microbes as their genomes became more compact. In this scenario, the exons were the building blocks of genes and, hence, of protein domains; this is known as the **introns early** model. The other idea, developed by Thomas Cavalier-Smith, was that introns were inserted into the coding regions of genes long after eukaryotes diverged from bacteria; this is the **introns late** hypothesis. The introns late view seems to be generally true for the classic introns of protein-coding genes but other kinds of introns may have much earlier origins (Rogers, 1990; Cavalier-Smith, 1991). Current interest in the history of introns is twofold: The distribution of the different classes of introns in various genomes may provide evidence for their degree of relatedness; and there may be some connection between protein-coding exons and the functional domains of protein molecules.

The protein domain is a fuzzy concept. On one hand, the term applies to specific regions of a protein that may interact with other macromolecules (for example, the DNA-binding **homeodomain** of the protein products of a class of regulatory genes implicated in the spatial organization of the body segments of animals and plants). Alternatively, a domain may be a block of amino acids that folds up into a coherent structure, often repeated within the molecule (Doolittle and Bork, 1993). In this sense, domains are motifs that have been copied and recopied, shuffled and reshuffled, and used and reused during the long course of protein evolution.

If classic introns appeared relatively late in the history of life, their migration into protein-coding genes may have promoted **domain shuffling** as a means of speeding up evolution. Suddenly, there was the possibility of duplicating, exchanging, or acquiring a domain by exon transfer, a process that is analogous to the **genetic recombination** that results from sexual reproduction. Each of these innovations—domain shuffling and sex—must have invigorated the evolution of life.

Another great source of evolutionary novelty comes from the duplication of whole genes. This has been a rare phenomenon in bacteria and other microbes compared with the abundant redundancy of the genomes of higher eukaryotes. For example, humans have more than a dozen genes for the oxygen transport molecule hemoglobin (also discussed in chapter 2). Some (the α_1 and α_2 genes) are recent duplicates set end to end on the same chromosome; others are ancient copies that have evolved apart and moved to different chromosomes (α_1/α_2 versus the β gene); still others are of intermediate age but are specialized to serve the fetus and embryo rather than the adult; some are extinct; but all are paired because about a billion years ago our ancestors moved from a **haploid** genome (n chromosomes) to a **diploid** one ($2n$ chromosomes).

A few genes have thousands of copies because their products are in great demand, (e.g., **ribosomal RNA (rRNA)** used in protein manufacture); when there are thousands of copies the genes are maintained as nearly identical copies by a process called **gene conversion**. However, most **multigene families** consist of far fewer copies; in these cases, gene conversion may operate for a while (human α_1 and α_2 hemoglobins are undergoing conversion at present), but the duplicated genes eventually become decoupled and evolve independently. A textbook example is the α and β hemoglobin genes of vertebrates. These genes were duplicated after the appearance of jawless fish 500 million years ago, but before the evolution of jaws that occurred about 50 million years later, so all living jawed vertebrates have a copy or copies of each gene (Dickerson and Geis, 1983). As pointed out by Jukes (1980; see also Chapter 2), the percentage differences in the amino acid sequences among pairs of α and β hemoglobins from a variety of living vertebrates average about 60% (62% ± 16%) (Runnegar, 1985), indicating that the rate of evolution has been approximately constant in all lines since the single hemoglobin gene was duplicated.

These are comparisons of **paralogous proteins** (α versus β); if the **orthologous proteins** (α versus α, β versus β) are compared among organisms, the percentage sequence differences tend to reflect evolutionary distances among the organisms being compared (discussed later).

The importance of gene duplications is that they allow repairs and modifications to be made while the vehicle is in motion. Copies of genes that originally served as backups or assistants may be co-opted for new functions as needs or opportunities arise. Thus, a great variety of cellular services may be performed by a single extended multigene family. Conversely, the history of gene duplications within and among organisms may be deduced from pairwise comparisons of the gene family sequences. This provides a way to explore deeply into the history of life—even beyond the common ancestor of all living organisms!

In the last quarter century, there has been much progress from the "one gene, one enzyme" reductionism of the heroic period of population genetics. However, there is a widespread intuition that "one genome, one organism" is the best way to characterize most living things, and that gene transfers among organisms occur rarely, if at all. How valid are these points of view?

Chimeric Life

Microbiologists now recognize about eight major groups (kingdoms) of living organisms: **Eubacteria** (true bacteria and cyanobacteria); **Archaebacteria** (methanogens, halobacteria, sulfobacteria), **Archezoa** (eukaryotes without organelles); **Protista** or **Protozoa** (various unicellular eukaryotes); **Chromista** (e.g., diatoms, brown algae, kelps); **Plantae** (red algae and green plants); **Fungi**; and **Animalia** (Cavalier-Smith, 1993). The last five kingdoms contain organisms that have **intracellular organelles**: **mitochondria** for aerobic respiration and **plastids** (chloroplasts, rhodoplasts, chromoplasts) for photosynthesis. Members of the remaining eukaryotic kingdom, Archezoa, lack these organelles and probably never had them.

Twenty-five years ago, Lynn Margulis and Peter H. Raven championed the nineteenth century idea that organelles are the remains of bacterial symbionts that invaded other microbes to create the ancestors of modern eukaryotic cells. This **endosymbiotic hypothesis** was strikingly confirmed when the DNA sequence of the maize chloroplast rRNA gene became available; the chloroplast gene is more similar to bacterial genes than it is to the equivalent gene in the maize nucleus (see Chapter 1 and Figure 1.14).

Early enthusiasm for the endosymbiont hypothesis led to the idea that there have been many separate endosymbiotic events in the history of life. Now that more molecular data are available, including complete sequences of some organelle genomes, the case for multiple endosymbioses has diminished (this argument applies only to transfers from bacteria to eukaryote not to subsequent transfers of organelles among eukaryotes). There are many genomic similarities among the plastids of green algae, green plants, red algae, and chromists that are best explained by an involved series of passes from host to host (Palmer, 1993; Figure 3.3).

The detail of the network is not important here, rather, the scale on which the network has operated and the difficulties that had to be faced. If the pathways shown in Figure 3.3 are approximately correct, the engine of photosynthesis emerges as a prize that has been invented once in the history of life and was carried from clade to clade as opportunities arose. There are some unexpected consequences: **RUBISCO** (ribulose-1,5-bisphosphate carboxylase/oxygenase), a key enzyme for photosynthesis, is, in higher plants, an assembly of eight large subunits coded by a chloroplast gene (**rbcL**) and eight small subunits coded by nuclear genes (**rbcS**). The rbcS gene was shifted to the nucleus along with other chloroplast genes after endosymbiosis, so its products now have to be reimported by the

FIGURE 3.3

Illustration of the relationships of kingdom-level groups of living organisms. The nuclear genomes of higher eukaryotes acquired some genes from the eubacterial symbionts that became the cellular organelles: Mitochondria and plastids (chloroplasts of green plants, rhodoplasts of red algae, and chromoplasts of brown algae and kelps). Arrows indicate likely pathways. Relationships are deduced mainly from molecular data.

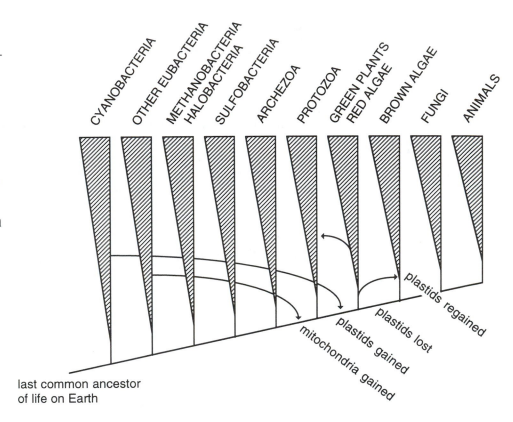

organelle to assemble the whole enzyme. This example is a microcosm of the chimeric nature of eukaryotic genomes.

Close encounters of cohabiting cells have probably been the principal causes of **lateral gene transfer**, but any organisms that live in proximity are vulnerable to invasion by foreign genes. In fact, it has been suggested that there may be a continuous flow of small amounts of genetic material among related species that live side by side. Regardless of whether this is true, it seems that lateral transfers of whole genes have been relatively rare events in the history of life (Smith et al., 1992).

MOLECULAR PALEONTOLOGY

Ancient DNA

In *Jurassic Park*, DNA obtained from Mesozoic amber was used to resurrect dinosaurs (Crichton, 1990). This delightful scenario may seem farfetched, but some insect DNA may have been amplified and sequenced from 130-million-year-old Lebanese amber (Cano et al., 1993). That insect DNA dated from the age of dinosaurs.

There have been several reports of DNA extracted from fossils tens of millions of years old (Pääbo, 1993). The critical technique is **amplification** of the tiny amounts of DNA left in the rocks by the **polymerase chain reaction (PCR)** (see Box 3.1 and Chapter 6). However, PCR is a two-edged sword; it will also amplify tiny amounts of modern DNA if any are present. Therefore, it is wise to be skeptical of all reports (Lindahl, 1993); to check research protocols for reproducibility and negative controls; and to look for corroborating evidence for the authenticity of ancient DNA.

When the DNA sequence obtained from an identified fossil is more similar to that of a living relative than to any other homologous sequence, the source of the DNA seems clear. For example, very different fossil leaves (magnolia and bald cypress) obtained from 17 to 20 million-year-old lake beds in Idaho, each yielded long sequences of the chloroplast *rbc*L gene that proved to be nearly identical to sequences from their closest living relatives, *Magnolia* and *Taxodium*, respectively (Golenberg et al., 1990; Soltis et al., 1992). Similarly, a partial *rbc*L sequence from an extinct species of *Hymenaea* (Fabaceae) found in 35 million-year-old Dominican amber was more like living species of the same genus rather than that of more distantly related plants (Poinar et al., 1993; Figure 3.4).

The *rbc*L gene was used as a target in each of these cases because its sequence is greatly conserved in different lines of evolution (Figure 3.5). It is therefore possible to design **PCR primers** (Box 3.1) that will work with a wide range of *rbc*L sequences. For example, the *rbc*L sequence of the extinct species of *Taxodium* was obtained using primers copied from the first 30 and last 30 DNA nucleotides of the maize *rbc*L sequence. This is impressive sequence conservation; the bald cypress (conifer) and maize (flowering plant) last shared a common ancestor more than 300 million years ago. Amazingly the *Taxodium* sequence obtained (1,320 contiguous, undamaged nucleotides) represents almost 1% of the chloroplast genome. As the sampling was random, it is likely that a substantial portion of the chloroplast genome remains intact; in principle, it should be possible to sequence the whole genome, synthesize it, and run it in a modern cell. This would not quite mimic *Jurassic Park*, but it demonstrates the change in expectations from a few years ago!

Bone and Shell Proteins

When W.T. Astbury discovered and named the alpha (α) structure (Linus Pauling's α-helix) in the protein keratin, he used a hair from the head of Mozart to obtain an x-ray diagram with characteristic reflections at 10Å and 5.1Å. It was immediately obvious that some proteins could survive in an essentially unmodified condition for long periods of time; this proved to be true for 33,000-year-old mammoth keratin from Alaska, which also yielded a clear 5.1Å signal (Gillespie, 1970).

The use of **antibodies** to seek out specific proteins forms the basis of **immunological** methods. Antibodies recognize the specific three-dimensional structure of a small region, an **antigenic site**, of the target protein. The structural integrity of such sites in **albumin** (a blood plasma protein) survives in mammoth and mastodon tissues, and there are reports of antigen-antibody reactions from fossils as old as 75 million years (Muyzer et al., 1992). However, the extraction and purification of intact protein molecules from fossilized bones and shells has proved a formidable task.

Because **biopolymers** (chains of variable components such as amino acids or sugars) participate in the formation of mineral skeletons, looking for them in fossil bones and shells makes sense. The idea that skeletal proteins and **polysaccharides** (polymers of sugars such as cellulose or starch) may be protected by the biominerals in which they occur was appreciated by Philip Abelson as early as 1954, and it is this principle that links molecular paleontology with a large body of recent research in biomineralization.

Broadly speaking, there seem to be two classes of proteins that participate in the production of mineral skeletons—those that form a structural framework and those that are attached to the framework in such a way as to control the site of deposition and the crystallography of the hard mineral component (Lowenstam and Weiner, 1989). The structural proteins are normally insoluble in water and rich in the amino acid **glycine**; the other kinds of proteins obtained by dissolving shell or bone in weak acids are rich in negatively charged amino acids (aspartic or glutamic acids) that are assumed to interact with the metal ions of the mineral crystallites. For example, in **vertebrate** bone, the structural framework

BOX 3.1

The Polymerase Chain Reaction (PCR)

The introduction of PCR in 1985 transformed molecular biology and had a revolutionary impact on evolutionary biology, paleoanthropology, biodiversity surveys, forensic science, and even legal issues. PCR works by amplifying DNA sequences between conserved or previously sequenced regions of the DNA.

The PCR chain reaction is explained in illustration on page 61. The target sequence (panel A) must be bounded by short stretches of known sequence; these may be known because the sequence of the target region has been sequenced previously or because the regions bounding the target may be predicted from existing knowledge of related genes. The related genes may be the other members of a multigene family within the same cell or, alternatively, copies of the same gene in other organisms (related genes in different organisms are called orthologs; related genes within the same organism are called paralogs).

DNA is usually found in the form of a double helix. Each of the two strands consists of a string of nucleotides A, T, G, and C. These nucleotides have the special characteristic that A will ordinarily only bind to T, and G will only bind to C. Hence, the string of nucleotides on one strand completely specifies the nucleotides on the other strand; for example, CCAGT must be matched by GGTCA. The two matching strands in the double helix are said to be complementary. Each strand of DNA has a "front" and a "back" end. The "front" end is called the 5' end (pronounced "5 prime"); the "back" end is called the 3' end. When DNA sequences are read by a cell, whether for replication or for protein synthesis, they are always read from 5' to 3'.

Before the PCR is initiated, short lengths of DNA complementary to the conserved regions upstream of the target sequence on both strands are synthesized in the laboratory (these oligonucleotides are called primers). The first step in the PCR reaction is the heating of the DNA from the organism or tissue sample of interest (the native DNA) in solution until the double helix "melts" into single strands (panel B). As the solution cools, crowds of primer molecules added to the solution out-compete the native strands in finding and pairing with their respective partners (panel C). Each of the primers that has attached itself to the native DNA is then lengthened in the downstream direction (5' to 3') by an enzyme called a polymerase. The polymerase adds free nucleotides, the building blocks of DNA that have been added to the solution, to match the sequence of bases in the exposed target DNA. The result are two double-stranded copies of the target sequence plus the primer regions (panel D); the rest of the native DNA is not amplified by the PCR reaction. Hence, there is a selective amplification of the desired sequence over all others present in the initial sample.

Each PCR cycle of melting, pairing, and copying doubles the number of molecules of the target sequence; twenty cycles produce about a million copies (2^{20}) and thirty cycles create more than a billion (2^{30}). As the reagents must withstand the high temperatures required to melt normal DNA, the DNA polymerases from heat-loving (thermophilic) bacteria have proved vital for this task (for example, *Taq* DNA polymerase from the hot vent microbe, *Thermus aquaticus*).

The PCR-amplified DNA has a known length and may therefore be isolated by gel electrophoresis, a technique that separates DNA strands by length (under an electrical field small strands migrate faster through the gel than long strands). The nucleotide sequence of the amplified DNA can be determined directly, although the amplified DNA is often cloned in the bacterium *Escherichia coli* in which it can be handled more easily. Because PCR is such an efficient way of amplifying target sequences, it may amplify an unintended source (for example, DNA from parasites rather than from their hosts may be amplified); this can be a problem if the DNA is in short supply (for example, in paleoanthropological and forensic applications) or if the nature of the target sequence is unpredictable (in phylogenetic studies of distantly related organisms). Corroborating evidence for the ultimate source of PCR-amplified DNA is always useful. For example, amplified sequences can often be compared with existing data bases to determine whether the amplified sequence belongs to the organism of interest, or to some contaminant.

REFERENCES

Erlich, H.A., Gelfand, D., and Sninsky, J.J. 1991. Recent advances in the polymerase chain reaction. *Science 252:* 1643–1651.

POLYMERASE CHAIN REACTION (PCR) ILLUSTRATED

A. Target DNA to be amplified

5' ——————————————— Target Sequence —————————————— 3'

AAACATGATCACTGGTGCAAGCCAGGCGGAACGTGGTGTGGGAATGAGCGCTGAAGGACAGACTAGAAACATATA
|||||||||||||||||||||||||||||||||||||||||||||||||||||||||||||||||||||||||||||
TTTGTACTAGTGACCACGTTCGGTCCGCCTTGCACCACACCCTTACTCGCGACTTCCTGTCTGATCTTTGTATAT

3'   5'

B. Melt DNA

AAACATGATCACTGGTGCAAGCCAGGCGGAACGTGGTGTGGGAATGAGCGCTGAAGGACAGACTAGAAACATATA
|||||||||||||||||||||||||||||||||||||||||||||||||||||||||||||||||||||||||||||

|||||||||||||||||||||||||||||||||||||||||||||||||||||||||||||||||||||||||||||
TTTGTACTAGTGACCACGTTCGGTCCGCCTTGCACCACACCCTTACTCGCGACTTCCTGTCTGATCTTTGTATAT

C. Anneal Primers

AAACATGATCACTGGTGCAAGCCAGGCGGAACGTGGTGTGGGAATGAGCGCTGAAGGACAGACTAGAAACATATA
|||||||||||||||||||||||||||||||||||||||||||||||||||||||||||||||||||||||||||||
                                                            TGTCTGATCTTTGTATAT
    5' primer   3' primer
AAACATGATCACTGGT
|||||||||||||||||||||||||||||||||||||||||||||||||||||||||||||||||||||||||||||
TTTGTACTAGTGACCACGTTCGGTCCGCCTTGCACCACACCCTTACTCGCGACTTCCTGTCTGATCTTTGTATAT

D. New Strands Synthesized

AAACATGATCACTGGTGCAAGCCAGGCGGAACGTGGTGTGGGAATGAGCGCTGAAGGACAGACTAGAAACATATA
|||||||||||||||||||||||||||||||||||||||||||||||||||||||||||||||||||||||||||||
TTTGTACTAGTGACCACGTTCGGTCCGCCTTGCACCACACCCTTACTCGCGACTTCCTGTCTGATCTTTGTATAT

AAACATGATCACTGGTGCAAGCCAGGCGGAACGTGGTGTGGGAATGAGCGCTGAAGGACAGACTAGAAACATATA
|||||||||||||||||||||||||||||||||||||||||||||||||||||||||||||||||||||||||||||
TTTGTACTAGTGACCACGTTCGGTCCGCCTTGCACCACACCCTTACTCGCGACTTCCTGTCTGATCTTTGTATAT

E. Re-Melt DNA . . .

formed from the protein **collagen** and a smaller molecule, **osteocalcin**, is believed to be implicated in the deposition of the mineral component (**hydroxyapatite**). Most osteocalcin molecules contain three residues of an unusual amino acid, **γ-carboxyglutamic acid** (known as Gla); each Gla residue has two negative charges that are thought to interact with positively charged calcium ions (Ca^{2+}) of the hydroxyapatite in the mineral lattice.

Bone collagen and osteocalcin are used as sources of carbon and nitrogen for carbon dating and dietary analysis. However, there is a hope that it will eventually be possible to extract and sequence proteins, such as osteocalcin, that are embedded in the bone and shell mineral during growth. So far, γ-carboxyglutamic acid has been reported from dinosaur remains up to 150 million years old (Muyzer et al., 1992) and a partial amino acid sequence of osteocalcin has been obtained from the bone of a 3,600-year-old New Zealand **moa** (Huq et al., 1990).

Biomarkers

The expression "molecular paleontology" seems to have been invented by Melvin Calvin when he gave the Bennett Lecture at Leicester University in 1968. Calvin and Geoffrey Eglinton had been using the new technique of **gas chromatography/mass spectrometry**

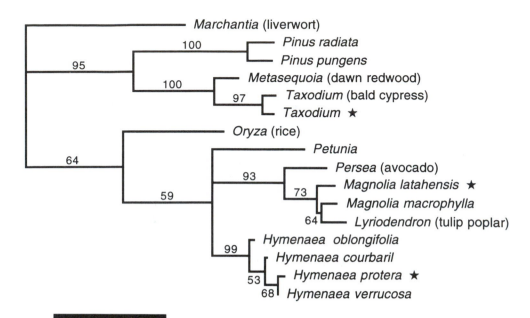

FIGURE 3.4

Evolutionary tree derived from full and partial *rbc*L gene sequences. The star (★) indicates gene sequences reported from three Tertiary fossil plants. Data analysis was done with PAUP 3.1.1 (Swofford, 1993) using the heuristic search option (informative sites only, random stepwise addition) for 2,000 bootstrapped replicates; the numbers represent the percentage of the replicates that supported overlying next node (one node with less than 50% support was collapsed to a trichotomy); missing data were coded as unknown. Data are from Golenberg et al. (1990), Soltis et al. (1992), and Poinar et al. (1993), plus additional sequences downloaded from Genbank using Genetics Computer Group (1991) software.

```
MARCHANTIA          CGTCTTGAAGATTTAAGAATTCCTCCAGCTTACACAAAAACTTTCCAAGGTCCTCCTCAT
PINUS RADIATA       CGTTTGGAAGATTTGCGGATTCCCCCTGCTTATTCCAAAACTTTTCAGGGTCCACCTCAT
PINUS PUNGENS       CGTTTGGAAGATTTGCGGATTCCCCCTGCTTATTCCAAAACATTTCAAGGTCCACCTCAT
METASEQUOIA         CGTCTGGAAGATTTACGAATTCCTCCTGCTTATTCAAAAACTTTCCAAGGACCACCACAT
TAXODIUM            CGTCTGGAAGATCTACGAATTCCTCCTGCTTATTCAAAAACTTTCCAAGGCCCACCACAT
TAXODIUM FOSSIL     CGTCTGGAAGATCTACGAATTCCTCCTGCTTATTCAAAAACTTTCCAAGGCCCACCACAT
ORYZA (RICE)        CGTCTGGAGGATCTGCGAATTCCCCCTACTTATTCAAAAACTTTCCAAGGTCCGCCTCAT
PETUNIA             CGTCTGGAAGATCTGCGAATCCCTCCTGCTTATGTTAAAACTTTCCAAGGGCCGCCTCAT
PERSEA (AVOCADO)    CGTCTGGAGGATCTGCGAATTCCTCCTGCTTATTCCAAAACTTTCCAAGGCCCGCCCCAT
MAGNOLIA FOSSIL     CGTCTGGAGGATCTGCGAATTCCTACTGCTTATGTCAAAACTTTCCAAGGCCCGCCTCAT
MAGNOLIA            CGTCTGGAGGATCTGCGAATTCCTACTGCTTATGTCAAAACTTTCCAAGGCCCGCCCCAT
LYRIODENDRON        CGTCTGGAAGATCTGCGAATTCCTCCTGCTTATATCAAAACTTTCCAAGGCCCGCCCCAT
                    ***1*3**3***1*31*3**3**31*31****3123*****3**3**3**3**3**3***
```

FIGURE 3.5

Samples of the set of aligned *rbc*L gene sequences analyzed for Figure 3.4. The lack of mismatches caused by insertions or deletions is typical of the whole *rbc*L gene. The sample begins at position 400 of the protein coding region and is 20 codons (60 nucleotides long); invariant nucleotides are indicated by asterisks. Note the diminishing frequency of substitutions in the third (3), first (1), and second (2) positions of the codons, respectively; overall, each of these positions is evolving at a different rate (see also Chapter 2).

(**GCMS**) to obtain and identify molecules of biological origin in sedimentary rocks (Eglinton and Calvin, 1967). By 1968 these chemical fossils were becoming known as **biological marker compounds**—now usually contracted to **biomarkers**—and their significance to the oil industry was beginning to be appreciated. Now, nearly thirty years later, the organic geochemistry of sedimentary basins has become a major enterprise. The importance of the field continues to grow as analytic techniques improve and the taxonomic origin and environmental significance of individual biomarkers become understood (Ourisson et al., 1984; Summons, 1988; Waples and Machihara, 1991).

In historical biology, the goal is to link particular biomarkers with their source organisms. This may require extensive biochemical surveys of living organisms, the recognition of *postmortem* modifications caused by burial and alteration, and the preparation of key compounds for standards that can be used to confirm the identity of the natural products. For example, **dinosteranes** are geologically modified compounds similar to **cholesterol** that are believed to originate in the cell membranes of **dinoflagellates** (photosynthetic protists that float in the ocean or occur as symbionts in other eukaryotes). The tough **cysts** of free-living dinoflagellates are important microfossils for dating rocks younger than about 200 million years (Helby et al., 1987). Like dinosaurs, undoubted **dinocysts** first appeared in the Triassic Period about 230 million years ago, despite the fact that the Dinoflagellata diverges deeply within trees constructed from rRNA genes (Cavalier-Smith, 1993), implying a billion-year-long history. This anomalous break in the fossil record of dinoflagellates is matched by the abrupt appearance of abundant dinosterane in Triassic sediments after a 250-million-year absence from preceding Paleozoic strata (Summons et al., 1992). We know that plankton diversity dropped suddenly some 600 million years ago; is this biomarker recording the ups and downs of dinoflagellate existence?

Biomarkers go back even further: Traces of **cholesteranes** (degraded cholesterol) extracted from 1.7-million-year-old sediments in northern Australia provide circumstantial evidence for the existence of eukaryotic cells at that time (Summons et al., 1988); and the isotopic compositions of prebiotic amino acids extracted from **carbonaceous meteorites** place constraints on how and where the primordial soup of prebiotic organic compounds might have formed as a precursor to life on Earth (Bada, 1991; Kerridge, 1991; Chyba and Sagan, 1992). Three alternatives are being examined: on Earth, within the solar nebula, or in interstellar space.

Generally speaking, the utility of molecular paleontology diminishes with the age of the samples. The obvious areas of growth are in historical disciplines in which time is measured in hours to years (forensic science) or thousands to millions of years (archeology, biodiversity surveys, global change research, paleoanthropology), rather than in the millions to billions of years of geology, planetology, and paleontology (Runnegar, 1985; Pääbo et al., 1989; Hagelberg et al., 1991). However, the meager data obtained with great difficulty from ancient sedimentary rocks and carbonaceous meteorites are of fundamental importance if we are to understand how life originated and diversified. In molecular paleontology, even the smallest clue may have great significance if its message can be detected, deciphered, and understood.

RECONSTRUCTING THE TREE OF LIFE

Stem Group Life

There is only one brand of life on Earth. All living things are descended from an organism that had DNA as its genetic material, used the common genetic code (Figure 3.1), and employed much of the molecular paraphernalia found in modern bacterial cells. This

last common ancestor of all living things founded the **crown group** of life (all descendants, living and extinct, of the last common ancestor). One of the great challenges of evolutionary biology is to understand the nature of the common ancestor and the history of life that preceded it. These were the organisms of life's **stem group** (organisms that preceded, coexisted with, and even followed the common ancestor but were never part of the crown group). Together with the crown group, they constitute the **total group** that began with the origin of life.

Life on Earth probably originated during the final stages of heavy bombardment of the inner solar system (4.0 billion to 4.4 billion years ago). As the flux of large objects hitting the Earth declined, a point was reached when the planet could be continuously inhabited. Even so, occasional energetic impacts may have boiled parts of the oceans, with the result that heat-tolerant (**thermophilic**) microorganisms should have had the best chance of surviving (Sleep et al., 1989).

One current speculation is that the last common ancestor of life on Earth was a **hyperthermophile** (an organism with an optimal temperature range of about 90°C). This idea has some support from trees built from rRNA genes (see Figure 1.14 in Chapter 1); most of the known (and aptly named) hyperthermophiles (for example, *Aquifex pyrophilus*, *Pyrococcus furiosus*, *Pyrodictium abyssi*, *Thermotoga thermarum*) lie near the root of the rRNA tree (Stetter, 1994). This observation has been extended to explain the origin of life itself, by the following logic: Hyperthermophiles are the most primitive organisms alive today; early Earth was more radioactive, and therefore hotter, than it is now; many modern hyperthermophiles are **chemosynthetic autotrophs** obtaining energy from the oxidation of hydrogen (H_2), sulfur (S^0), or sulphide (S^{2-}); life began on iron sulfide crystals where energetic surface reactions built organic compounds directly from carbon dioxide (CO_2) and water; therefore, the cradle of life was a high-temperature, sulfur-rich environment like the hot vents and black smokers on the modern sea floor; Q.E.D.

Nonsense. First, the organisms at the base of the rRNA tree belong to the crown group of life; they may postdate the origin of life by a billion years. Although Schopf (1993) identified some 3.5 billion-year-old fossils as probable cyanobacteria (see Chapter 4), they do not necessarily belong to life's crown group. For example, some early fossil birds such as *Archaeopteryx* lived before the last common ancestor of all living birds. These archaic birds are members of the bird stem group. The same may be true for Schopf's fossils; they may be stem group cyanobacteria (or even another kind of extinct organism). Second, the hyperthermophiles belong to two very distinct bacterial kingdoms that had evolved far apart (*Aquifex* and *Thermotoga* are true bacteria, *Pyrococcus* and *Pyrodictium* are archaebacteria). Were all organisms on the planet hyperthermophiles living in near-boiling oceans? Possibly. If the rRNA tree is correct, it implies that heat generated by an ocean-boiling CO_2 greenhouse (Lowe, 1994) or a late large impact may have filtered the biosphere of all but the extreme thermophiles. However, a likely alternative is that the rRNA tree is incorrect, perhaps because it is biased by the composition of the DNA; thermophile DNA tends to be GC-rich because it is more stable at high temperatures (due to the presence of three hydrogen bonds between G-C pairs instead of two between A-T pairs).

Although the genomes of all living species converge on the genome of the last common ancestor as we move backward in time, this is not the final obstacle to historical research. We can look beyond the genome of the last common ancestor because that genome was, at least in part, composed of sets of copies of preexisting genes (multigene families). Their histories, which may be deduced by sequence comparisons, take us back beyond the last common ancestor of life on Earth.

For example, **elongation factors** are conserved proteins that are involved in protein synthesis in all cells. One kind of elongation factor, EF-1 (EF-Tu or EF1α), assists in the binding of activated tRNAs to ribosomes, a step in protein production; another kind of elongation factor, EF-2 (EF-G or EF-2), is involved in the transfer of amino acids from activated tRNAs to the nascent protein. These two kinds of molecules are paralogous, having been produced by a gene duplication in the line leading to the common ancestor of modern life. Therefore, we can conclude that before the time of that gene duplication, both functions (binding and transfer) were carried out by a single elongation factor. No living organism operates in this fashion; it is a property of some members of the unobservable stem group.

Rooting the Tree of Life

A crucial step in constructing an evolutionary tree is to determine which branch point in the tree represents the common ancestor of all of the branch tips; this is called **rooting** the tree. Finding the root generally requires an **outgroup**, a group that branched off before the common ancestor of the **ingroup** (the organisms under study). What is the outgroup to life? Rocks? The ingenious answer is that genes duplicated before the origin of the crown group may be used to find the root of the Tree of Life (Iwabe et al., 1989). Each of a pair of paralogous genes acts as an outgroup to the other; for example, trees constructed from unambiguously aligned, conserved parts of EF-1 and EF-2 sequences (about 130 amino acids in length) may be rooted at the duplication event. When this is done, the topology of the EF-1 side of the tree is broadly similar to the EF-2 side (Figure 3.6). In particular, the archaebacteria (*Sulfolobus, Thermoplasma, Methanococcus,* and *Halobacterium*) group with the eukaryotes (*Homo* and *Drosophila*), not with the true bacteria (*Thermus* and *Escherichia coli*) on both sides of the tree. A similar solution was obtained with ATPase sequences, but glutamate dehydrogenase sequence comparisons gave ambiguous results. Thus, the final answer may not yet be known.

Origin of the Eukaryotes

Most modern eukaryotes are sizable, complex organisms composed of many different kinds of cells, each containing a nucleus, cytoskeleton, and organelles (mitochondria and

FIGURE 3.6

Rooting the Tree of Life. Tree constructed from greatly conserved parts of the amino acid sequences of elongation factor proteins 1 (EF-Tu or EF1a) and 2 (EF-G or EF-2). The genes for EF-1 and EF-2 arose from the duplication of a single ancestral elongation factor gene in the line leading to the last common ancestor of all living life. Therefore, EF-2 sequences may act as outgroups for EF-1 sequences (and vice versa). On both sides of this tree, the archaebacteria (*Sulfolobus, Thermoplasma, Methanococcus,* and *Halobacterium*) are more closely related to the eukaryotes (*Homo* and *Drosophila*) than they are to the eubacteria (*Escherichia coli* and *Thermus*). The procedures used to construct this tree are described in Runnegar (1994).

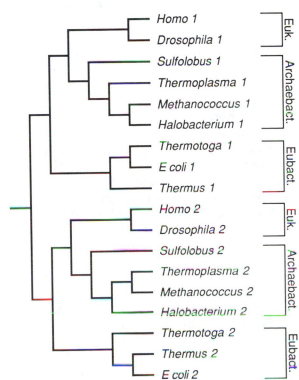

plastids) of endosymbiotic derivation. This remarkable complexity was not achieved instantaneously, so there is no unique origin of the eukaroytic cell. Instead, the characteristics of modern eukaryotes were evolved and assembled over a long period of time that began in the **Archean** period of Earth history (> 2.5 billion years ago).

As we have seen, the closest prokaryotic relatives of eukaryotes are probably the archaebacteria. However, archaebacteria are a mixed bag of extremophiles—organisms that inhabit extreme environments (hot, anoxic, saline, sulfurous, acidic, and so on). Are certain kinds of archaebacteria our closest bacterial relatives?

Trees obtained originally from rRNA sequence comparisons suggested that the Archaebacteria is a **monophyletic clade** (all descendants of a common ancestor) (Woese, 1987). Other analyses hinted that the group might be **paraphyletic** (only some descendants of a common ancestor) (Lake, 1988) but the debate could not be resolved, presumably because of ambiguities from homoplasy in the DNA sequences. What was needed was something more complex: A character that reveals relationships and yet is sufficiently complicated that the probability that it evolved twice is vanishingly small (see Chapter 1 for a discussion of the value of rare events in phylogeny reconstruction). Such a character was found in EF-1 sequences; it is an insert of 10 amino acids that is confined to eukaryotes and archaebacteria known as **eocytes** (sulfobacteria), but is absent from other archaebacteria, true bacteria, and from *all EF-2 sequences* (Rivera and Lake, 1992). Once again, the EF-2 outgroup (see Figure 3.6) helps show the way evolution has occurred. The conclusion seems inescapable: Eocytes such as *Sulfolobus* (see Figures 3.6 and 3.7) are the surviving **sister group** of the eukaryotes (see Figure 3.3).

Cambrian Explosion of Multicellular Life

Suddenly, after 4 billion years of Earth history, complex organisms burst into the fossil record and created the Cambrian explosion of multicellular life. Ever since Darwin, people have been puzzled by the abruptness of this event and the lack of precursors in the preceding Precambrian era. There are two endmember explanations. The null hypothesis is that complex life suddenly blossomed from various unicellular ancestors at the close of the Precambrian era. The alternative is that complex life has a cryptic, multimillion-year-long, Precambrian prehistory not seen in the fossil record (Runnegar, 1982).

Molecular methods may help resolve this dilemma. If all of the animal phyla originated and diversified within about 10 million years of the beginning of the Cambrian period, their rRNA gene sequences should have started to drift apart at roughly the same time. In other words, the race to the present began within earshot of the Cambrian explosion. Consequently, there should be no way to obtain the correct relationships by sequence comparisons among the various animal phyla. Sponges and snails, corals and urchins, flatworms and fish, all ought to have rRNA genes that are equally distant from each other (measured as percentage differences in pairs of sequences; any variability in the data should have a bell-shaped distribution around a mean value). However, if the observed distances can be used to derive previously postulated relationships among the animal phyla, the race cannot have been fair; some participants must have started before others (Figure 3.8). Thus, it seems clear that some animals (**metazoans**) had a substantial Precambrian prehistory.

Although the broad framework of the metazoan tree has been known from comparative anatomy, many important details remain unclear. Field et al. (1988) made a heroic attack on this problem by obtaining partial rRNA sequences from a small Noah's ark of invertebrate species. Although their study has been eclipsed by more recent work, it has had two long-lasting benefits: Wide sampling and contaminant-free sampling. Most invertebrates are infested with a wide range of microorganisms (intracellular symbionts, endoparasites, gut-inhabitants, and so forth). The Field et al. sequences were obtained from

```
                                        PRIMERAMPLIFIED BY RIVERA & LAKEPRIMER
HUMAN 1                          GHRDFIKNMITGTSQADCAVLIVAAGVGEFEAGISKNGQTREH
FRUIT FLY 1                      GHRDFIKNMITGTSQADCAVQIDAAGTGEFEAGISKNDQTREH
YEAST 1                          GHRDFIKNMITGTSQADCAILIIAGGVGEFEAGISKDGQTREH
TOMATO 1                         GHRDFIKNMITGTSQADCAVLIIDSTTGGFEAGISKDGQTREH
PLASMODIUM 1 (MALARIAL PARASITE) GHKDFIKNMITGTSQADVALLVVPADVGGFDGAFSKEGQTKEH
GIARDIA 1 (GUT PARASITE)         GHRDFIKNMITGTSQADVAILVVAAGQGEFEAGISKDGQTREH
SULFOLOBUS 1 (ARCHAEBACTERIUM)   GHRDFVKNMITGASQADAAILVVSAKKGEYEAGMSAEGQTREHI
HALOBACTERIUM 1 (ARCHAEBACTERIUM) GHRDFVKNMITGASQADNAVLVVAADDG     VQPQTQEH
SPIRULINA 1 (CYANOBACTERIUM)     GHADYVKNMITGAAQMDGAILVVSAADG     PMPQTREH
EUGLENA CHLOROPLAST 1            GHADYVKNMITGAAQMDGAILVVSAADG     PMPQTKEH
E.COLI 1 (BACTERIUM)             GHADYVKNMITGAAQMDGAILVVAATDG     PMPQTREH
THERMOTOGA 1 (BACTERIUM)         GHADYIKNMITGAAQMDGAILVVAATDG     PMPQTREH
THERMUS 1 (BACTERIUM)            GHADYIKNMITGAAQMDGAILVVSAADG     PMPQTREH
YEAST MITOCHONDRION 1            GHADYIKNMITGAAQMDGAIIVVAATDG     QMPQTREH
HUMAN 2                          GHVDFSSEVTAALRVTDGALVVVDCVSG     VCVQT ET
FRUIT FLY 2                      GHVDFSSEVTAALRVTDGALVVVDCVSG     VCVQT ET
SLIME MOLD 2                     GHVDFSSEVTAALRVTDGALVVIDCVEG     VCVQT ET
GREEN ALGA 2                     GHVDFSSEVTAALRITDGALVVVDCIEG     VCVQT ET
SULFOLOBUS 2                     GHVDFSGRVTRSLRVLDGSIVVIDAVEG     IMTQT ET
HALOBACTERIUM 2                  GHVDFGGDVTRAMRAVDGALVVVDAVEG     AMPQT ET
E.COLI 2                         GHVDFTIEVERSMRVLDGAVMVYCAVGG     VQPQS ET
THERMOTOGA 2                     GHVDFTAEVERALRVLDGAIRVFDATAG     VEPQS ET
THERMUS 2                        GHVDFTIEVERSMRVLDGAIVVFDSSQG     VEPQS ET
YEAST MITOCHONDRION 2            GHIDFTIEVERALRVLDGAVLVVCAVSG     VQSQT VT
                                 ANCESTRAL CONDITION:            •••
                                 10 AMINO ACID REPLACEMENT:   1234567890
```

FIGURE 3.7

Amino acid sequences of parts of the elongation factor proteins. These sequences were used by Rivera and Lake (1992) to test the hypothesis that eocytes (sulfobacteria) are the sister group of the eukaryotes (see Figure 3.3). Greatly conserved regions (KNMITG and QTREH, indicated on figure under "PRIMER") provided the knowledge needed to design PCR primers (see Box 3.1) that could be used to amplify only that part of the molecule that contains historically significant information (the stretch labeled "Amplified by Rivera & Lake"). A proline-rich (P-rich) section of three amino acids (the original condition, indicated as ···) was replaced by a stretch of ten different amino acids (1234567890) in eukaryotes (*Homo* to the gut parasite *Giardia*) and in eocyte archaebacteria (represented by *Sulfolobus*), but not in other archaebacteria (represented by *Halobacterium*), in eubacteria, in the organelles derived from them (chloroplasts and mitochondria), or in any EF-2 protein.

cellular rRNA without PCR amplification (Box 3.1) and are, therefore, likely to have come from the animals named rather than the associated microorganisms; PCR-amplified sequences may now be checked against the Field et al. standards.

The post-Field et al. rise in rRNA sequencing has rejuvenated research into invertebrate relationships. It is now possible to propose hypotheses based on anatomical observations and to test those hypotheses with independent molecular data sets. There are two pleasing outcomes of this work: rRNA sequence comparisons have supported already robust parts of the metazoan tree (confirming the value of the method), and they are beginning to settle some old controversies of invertebrate evolution (Figure 3.8). Briefly, animals (metazoans) are monophyletic and more closely related to fungi than to plants (Wainwright et al., 1993); **sponges** and **cnidarians** (corals, sea anemones, jellyfish) lie at the base of the metazoan tree; **flatworms** come in next, followed by **nematodes**; **chordates** (vertebrates, amphioxus, tunicates) group with **echinoderms** (urchins and starfish), as indicated by embryological ev-

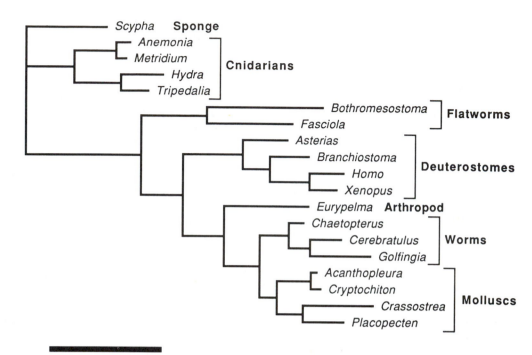

FIGURE 3.8

Relationships among animal phyla inferred from comparisons of pairwise differences among small subunit rRNA gene sequences. Broadly speaking, the tree displays the same groupings and relationships as those inferred from comparative anatomy and embryology: Four cnidarians (*Anemonia*, *Metridium*, *Hydra*, and *Tripedalia*) lie closest to the user-defined root (the sponge *Scypha*); two flatworms (*Bothromesostoma* and *Fasciola*) lie between the Cnidaria and the higher Bilateria; an echinoderm, *Asterias*, groups with the other deuterostomes (*Branchiostoma* [amphioxus], *Homo*, and *Xenopus*); an arthropod (the spider, *Eurypelma*) lies above the deuterostomes and outside the worms (*Chaetopterus*, *Cerebratulus*, *Golfingia*) and molluscs (*Acanthopleura*, *Cryptochiton*, *Crassostrea*, *Placopecten*). If all of these phyla had originated simultaneously in the Cambrian Explosion, this tree could not be derived from distance data. Sequences were aligned by eye and analyzed using the Dnadist and Neighbor options of PHYLIP 3.5c (Felsenstein, 1993).

idence; and most of the remaining invertebrates (**arthropods**, **molluscs**, various worm phyla) converge on an unresolved starburst that may be a molecular expression of the Cambrian explosion. In other words, there was a major radiation of segmented animals at or near the Precambrian–Cambrian boundary, approximately 545 million years ago.

CLOCKING THE EVOLUTION OF LIFE

Is it possible to date significant events in the history of life by using molecular and other techniques? Three components must be kept in mind: Fossils, divergence times, and isotopic dates. They rarely correspond and the relative ages of each have to be inferred. Dating the Tree of Life is a challenging task.

Isotopic Dates and the Fossil Record

Some of the best dates are from the oldest rocks. Bomb uranium (^{235}U) was fairly common millions of years ago, so its **radioactive decay product** (^{207}Pb) is easily measured in fossiliferous rocks. When this measurement is combined with measurements of lead (^{206}U) from reactor uranium (^{238}U), it is possible to calculate an exact age. Thus, some of the oldest fossils are from rocks that are known to be less than 3,471 $\pm$ 3 millions of years old but more than 3,458 $\pm$ 2 millions of years old (Schopf, 1993). It is not known whether these microfossils are the remains of crown group organisms.

The oldest convincing crown group fossils come from rocks that are about 2 billion years old, but the last common ancestor of living life lived long before that, because both methane-generating archaebacteria (**methanogens**) and eubacteria have left their traces in Archean rocks that are about 2.7 billion years old (Runnegar, 1994). The first three branches of the Tree of Life shown in Figure 3.3 must be older than 2.7 billion years by as yet unknown lengths of time.

A key tie point for calibrating the Tree of Life may come from the recent discovery of a probable eukaryotic **alga** called *Grypania* in Michigan iron ores that are about 2 billion years old (Han and Runnegar, 1992). This fossil implies that life had reached node 6 of the Tree of Life (Figure 3.3) before 2 billion years ago. If this is true, there is an enormous amount of time (1.5 billion years) between the origin of complex eukaryotic cells and the Cambrian explosion.

Molecular Clockwork

The observed similarity between any pair of correctly aligned protein or DNA sequences diminishes roughly in proportion to the evolutionary distance that separates the pair (Figure 3.9). This apparently regular decline in pairwise similarity, first observed in aligned protein sequences, led Emile Zuckerkandl and Linus Pauling to suggest that the rates of change may have been uniform enough to be used as a **molecular clock** (Zuckerkandl, 1987; Runnegar, 1991). This fruitful idea has been applied to a variety of problems in historical biology, but it has also been roundly criticized, both for the unverified assumptions on which it is based and for the way it performs in cases in which molecular dates may be compared with those derived from the fossil record. There are also internal tests of consistency that suggest that the molecular clock is frequently imprecise and unreliable.

Nevertheless, molecular clocks may provide valuable insights. Alan Wilson and Vincent Sarich's aggressive use of immunological data to estimate that the divergence between man and ape occurred as recently as about 5 million years ago overthrew established paleontological wisdom that preferred a much older time of separation (Sarich and Wilson, 1973). As is often the case in historical sciences, knowing something imprecisely is much better than knowing nothing. For example, there is now considerable evidence that human molecular clocks run more slowly than their rodent counterparts (Gibbs and Dugaiczyk, 1994). Why this should be so remains an important question for molecular and morphological biology.

DEDICATION

The 1994 CSEOL Symposium was held on March 18, exactly ten years after the premature death of Bill Schopf's older brother, Thomas J.M. Schopf. At the time of his death, Tom Schopf was working hard to build bridges from the earth sciences to molecular biology. This contribution to *Evolution and the Molecular Revolution* is dedicated to his memory.

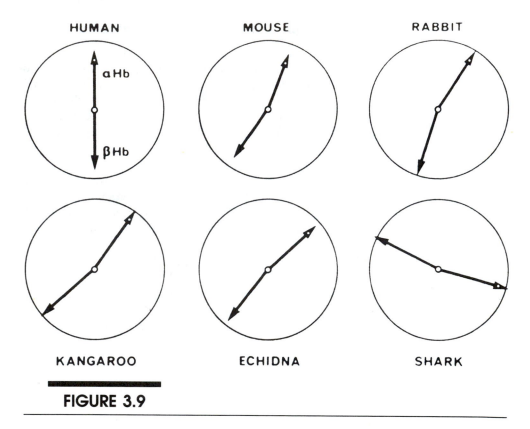

FIGURE 3.9

Molecular clocks. Illustration of how observed differences in the amino acid sequences of vertebrate hemoglobins may be used as rough measures of time. The diameter of the face of the clocks represents the average difference (62%) between the α and β hemoglobins of living vertebrates. The length of the hands corresponds to the observed difference for each animal analyzed. If hands representing the human α and β hemoglobins are placed at 12 and 6 o'clock, and the comparative sequence differences in other hemoglobins are shown as the clockwise distance away from the human position, the clocks show three things: (1) The sequence difference between α and β hemoglobins is similar in all six animals (the hands are roughly equal in length). (2) Both kinds of hemoglobin depart by a roughly equal amount from their human counterparts (the hands remain approximately 180° apart). (3) The amount of difference corresponds to the evolutionary distance from humans (the greater the evolutionary distance the further the hands have rotated, with shark showing the greatest rotation). (Reproduced from Runnegar, 1982, with permission.)

REFERENCES

Bada, J.L. 1991. Amino acid cosmochemistry. *Phil. Trans. Roy. Soc. Lond.* (B) *333:* 349–358.

Cavalier-Smith, T. 1991. Intron phylogeny: A new hypothesis. *Trends Gene. 7:* 145–148.

Cavalier-Smith, T. 1993. Kingdom protozoa and its 18 phyla. *Microb. Rev. 57:* 953–994.

Cano, R.J., Poinar, H.N., Pieniazek, N.J., Acra, A., and Poinar, G. 1993. Amplification and sequencing of DNA from a 120–135-million-year-old weevil. *Nature 363:* 536–538.

Chyba, C., and Sagan, C. 1992. Endogenous production, exogenous delivery and impact-shock synthesis of organic molecules: An inventory for the origins of life. *Nature 355:* 125–132.

Crichton, M. 1990. *Jurassic Park* (New York: Knopf).

Dickerson, R.E., and Geis, I. 1983. *Hemoglobin: Structure, function, evolution, and pathology* (Menlo Park: Benjamin/Cummings).

Doolittle, R.F., and Bork, P. 1993. Evolutionary mobile modules in proteins. *Scientific American 269*(4): 50–56.

Eglinton, G., and Calvin, M. 1967. Chemical fossils. *Scientific American 216*(1): 32–43.

Felsenstein, J. 1993. *PHYLIP (Phylogeny Inference Package) Version 3.5c.* Joseph Felsenstein and the University of Washington.

Field, K.G., Olsen, G.J., Lane, D.J., Giovannoni, S.J., Ghiselin, M.T., Raff, E.C., Pace, N.R., and Raff, R.A. 1988. Molecular phylogeny of the animal kingdom. *Science 239:* 748–753.

Genetics Computer Group. 1991. *Program Manual for the GCG Package, Version 7* (Madison, WI: Genetics Computer Group).

Gibbs, P.E., and Dugaiczyk, A. 1994. Reading the molecular clock from the decay of internal symmetry of a gene. *Proc. Natl. Acad. Sci. USA 91:* 3413–3417.

Gillespie, J.M. 1970. Mammoth hair: Stability of a-keratin structure and constituent proteins. *Science 170:* 110–112.

Golenberg, E.M., Giannasi, D.E., Clegg, M.T., Smiley, C.J., Durbin, M., Henderson, D., and Zurawski, G. 1990. *Nature 344:* 656–658.

Hagelberg, E., Gray, I.C., and Jeffreys, A.J. 1991. Identification of the skeletal remains of a murder victim by DNA analysis. *Nature 352:* 427–429.

Han, T.M., and Runnegar, B. 1992. Megascopic algae from the 2.1-billion-year-old Negaunee Iron Formation, Michigan. *Science 257:* 232–235.

Helby, R., Morgan, R., and Partridge, A.D. 1987. A palynological zonation of the Australian Mesozoic. *Mem. Ass. Australas. Palaeontols. 4:* 1–94.

Hofmann, H.J. 1976. Precambrian microflora, Belcher Islands, Canada: Significance and systematics. *J. Paleont. 50* (6): 1040–1073.

Huq, N.L., Tseng, A., and Chapman, G.E. 1990. Partial amino acid sequence of osteocalcin from an extinct species of ratite bird. *Biochem. Internat. 21:* 491–496.

Iwabe, N., Kuma, K., Hasegawa, M., Osawa, S. and Miyata, T. 1989. Evolutionary relationship of archaebacteria, eubacteria, and eukaryotes inferred from phylogenetic trees of duplicated genes. *Proc. Natl. Acad. Sci. USA 86:* 9355–9359.

Jukes, T.H. 1980. Silent nucleotide substitutions and the molecular evolutionary clock. *Science 210:* 973–978.

Kerridge, J.F. 1991. A note on the prebiotic synthesis of organic acids in carbonaceous meteorites. *Orig. Life Evol. Bios. 21:* 19–29.

Knoll, A.H., Strother, P.K., and Rossi, S. 1988. Distribution of microfossils from the lower Proterozoic Duck Creek Dolomite, Western Australia. *Precamb. Res. 38:* 257–279.

Lake, J.A. 1988. Origin of the eukaryotic nucleus determined by rate-invariant analysis of rRNA sequences. *Nature 331:* 184–186.

Lindahl, T. 1993. Instability and decay of the primary structure of DNA. *Nature 362:* 709–715.

Lowe, D.R. 1994. Early environments: Constraints and opportunities for early evolution. In: S. Bengtson (Ed.), *Early Life on Earth. Nobel Symposium 84* (New York: Columbia Univ. Press), pp. 24–35.

Lowenstam, H. A., and Weiner, S. 1989. *On Biomineralization* (Oxford: Oxford Univ. Press).

Muyzer, G., Sandberg, P., Knapen, M.H.J., Vermeer, C., Collins, M., and Westbroek, P. 1992. Preservation of the bone protein osteocalcin in dinosaurs. *Geology 20:* 871–874.

Ourisson, G., Albrecht, P., and Rohmer, M. 1984. The microbial origin of fossil fuels. *Scientific American 251* (2): 34–41.

Pääbo, S., Higuchi, R. G., and Wilson, A.C. 1989. Ancient DNA and the polymerase chain reaction. The emerging field of molecular archaeology. *J. Biol. Chem. 264:* 9709–9712.

Palmer, J.D. 1993. A genetic rainbow of plastids. *Nature 364:* 762–763.

Poinar, H.N., Cano, R.J., and Poinar, G.O. 1993. DNA from an extinct plant. *Nature 363:* 677.

Rivera, M.C., and Lake, J.A. 1992. Evidence that eukaryotes and eocyte prokaryotes are immediate relatives. *Science 257:* 74–76.

Rogers, J.H. 1990. The role of introns in evolution. *FEBS Letters 268:* 339–343.

Runnegar, B. 1982. The Cambrian explosion: Animals or fossils? *J. Geol. Soc. Aust. 29:* 395–411.

Runnegar, B. 1985. Molecular palaeontology. *Palaeont. 29:* 1–24.

Runnegar, B. 1991. Nucleic acid and protein clocks. *Phil. Trans. Roy. Soc. Lond.* (B) *333:* 391–397.

Runnegar, B. 1994. Proterozoic eukaryotes: Evidence from biology and geology. In: S. Bengtson (Ed.), *Early Life on Earth. Nobel Symposium* 84 (New York: Columbia Univ. Press), pp. 287–297.

Sarich, V.M., and Wilson, A.C. 1973. Generation time and genomic evolution in primates. *Science 179:* 1144–1147.

Schopf, J.W. 1993. Microfossils of the early Archean Apex chert: New evidence of the antiquity of life. *Science 260:* 640–646.

Sleep, N.H., Zahnel, K.J., Kasting, J.F., and Morowitz, H.J. 1989. Annihilation of ecosystems by large asteroid impacts on the early Earth. *Nature 342:* 139–142.

Smith, M.W., Feng, D., and Doolittle, R.F. 1992. Evolution by acquisition: The case for horizontal gene transfers. *Trends Biochem. Sci. 176:* 489–493.

Soltis, P.S., Soltis, D.E., and Smiley, C.J. 1992. An *rbc*L sequence from a Miocene *Taxodium* (bald cypress). *Proc. Natl. Acad. Sci. USA 89:* 449–451.

Stetter, K.O. 1994. The lesson of Archaebacteria. In: S. Bengtson (Ed.), *Early Life on Earth. Nobel Symposium* 84 (New York: Columbia Univ. Press), pp. 143–151.

Summons, R.E. 1988. Biomarkers: Molecular fossils. In: B. Runnegar and J.W. Schopf (Eds.), *Molecular Evolution and the Fossil Record. Short Courses in Paleontology 1:* 98–113.

Summons, R.E., Powell, T.G., and Boreham, C.J. 1988. Petroleum geology and geochemistry of the middle Proterozoic McArthur Basin, northern Australia: III. Composition of extractable hydrocarbons. *Geochim. Cosmochim. Acta 52:* 1747–1763.

Summons, R.E., Thomas, J., Maxwell, J.R., and Boreham, C.J. 1992. Secular and environmental constraints on the occurrence of dinosterane in sediments. *Geochim. Cosmochim. Acta 56:* 2437–2444.

Swofford, D.L. 1993. *PAUP: Phylogenetic Analysis Using Parsimony, version 3.1* (Champaign, IL: Illinois Natural History Survey), computer program.

Wainwright, P.O., Hinkle, G., Sogin, M.L., and Stickel, S.K. 1993. Monophyletic origins of the Metazoa: An evolutionary link with Fungi. *Science 260:* 340–342.

Waldrop, M. M. 1992. *Complexity* (New York: Simon & Schuster).

Waples, D.W., and Machihara, T. 1991. Biomarkers for geologists. *AAPG Meth. Explor. 9:* 1–91.

Woese, C.R. 1987. Bacterial evolution. *Microbiological Reviews 51:* 221–271.

Zuckerkandl, E. 1987. On the molecular evolutionary clock. *J. Mol. Evol. 26:* 34–46.

CHAPTER 4

METABOLIC MEMORIES OF EARTH'S EARLIEST BIOSPHERE

■

J. William Schopf*

LIVING SYSTEMS ARE SIMPLE!

The Components and Elements of Life

Living systems are not complicated. In fact, all forms of life are made of a surprisingly small number of chemical ingredients. You and all the rest of the living world—even a head of lettuce—are made mostly of water. Why? Because living systems originated in aqueous surroundings; water was the cradle of life. As a result, almost all the chemistry of life is performed in water, the water that makes up the liquid medium, the **cytosol**, of living cells. Our watery composition simply reflects our evolutionary heritage.

Look at the elements of life listed in Figure 4.1. There are only a very few: **CHON**—carbon, hydrogen, oxygen, and nitrogen—and sometimes P (phosphorus) and S (sulfur). Why CHON? Why not titanium, gold, krypton, and thulium, or some other exotic mixture? Again, the answer is simple. Life is composed of CHON chiefly because these four elements are both widespread and plentiful. In fact, carbon, hydrogen, oxygen, and nitrogen are four of the five most abundant elements in the entire universe—there was a lot of CHON around when life got started. Moreover, all these elements are able to combine with one another to form small sturdy molecules (such as methane, CH_4; carbon dioxide, CO_2; ammonia, NH_3; and many others), compounds that dissolve in water and can thus play an active role in the chemistry of life. (The fifth abundant element, helium, is inert and nonreactive—a great gas for balloons, but an element that is unable to combine with any others to form robust chemical compounds.)

The Chemistry of Life: Monomers, Polymers, and Enzymes

Over 99% of the **biosphere** (the sum total of *all* living systems) is composed of CHON. That seems simple enough, but what about the mind-boggling array of biological com-

---

*IGPP Center for the Study of Evolution and the Origin of Life, Department of Earth and Space Sciences, and Molecular Biology Institute, University of California, Los Angeles 90095.

ELEMENTAL COMPOSITION			COMPONENT OF LIFE	COMPOSITION OF REPRESENTATIVE ORGANISMS								
CARBON OXYGEN PHOSPHORUS HYDROGEN NITROGEN SULFUR				Lettuce	Celery	Mushroom	Oyster	Codfish	Bacterium	Cow	Chicken	Pig
H O			WATER (Universal Solvent)	95%	94%	90%	88%	83%	75%	74%	66%	57%
C H O N		S	PROTEIN (Enzyme Catalysts)	1.3	1.4	3.6	6.0	12.0	17.5	19.6	21.2	20.1
C H O			FAT (Energy Storage)	0.4	0.4	0.4	1.5	3.5	2.5	4.2	11.0	20.2
C H O			CARBOHYDRATE (Cell Walls)	2.1	3.0	5.1	2.4	0	1.3	0	0	0
C H O N P			DNA, RNA, ATP (Genes, Energy)	1.2	1.2	0.9	2.1	1.5	3.7	2.2	1.8	2.7

FIGURE 4.1

The chemical components of living systems. In addition to CHON and P and S, numerous other chemical elements (magnesium, iron, copper, and cobalt, for example) are also required by living systems, but only in trace amounts.

pounds made up of these four elements? Some biological molecules are huge, composed of thousands, even millions of individual atoms (such as the DNA of chromosomes), but they nevertheless are far simpler than one might imagine, because virtually *all large biological compounds are composed of an orderly sequence of much smaller and much simpler subunits*. For example, **cellulose**, the material that makes up plant cell walls, the most abundant biological compound on Earth, accounts for more than half of all carbon in the entire biosphere. Cellulose is an exceedingly long molecule, made up of thousands of carbon, hydrogen, and oxygen atoms. However, as shown in Figure 4.2, cellulose is also a simple molecule, composed of an orderly, repeating sequence of small subunits, each of which contains only six carbon atoms. The repeating subunit, or **monomer**, is **glucose**, a six-carbon sugar; and the sugar monomers are linked together like beads on a string to form a **polymer**, the simple but very long cellulose molecule.

Many such polymers occur in nature. **Natural rubber,** for example, is composed of a repeating linear beads-on-a-string assembly of thousands of short five-carbon (C5) **isoprene** monomers, also shown in Figure 4.2. Like cellulose, polymers made of isoprene subunits are abundant in the living world. Two isoprene monomers can be linked together (see Figure 4.2) to form **geranyl**, a ten-carbon (C10) polymer that is important as a building block of testosterone, estrogen, and related human hormones. A third isoprene monomer can be linked to geranyl to form **farnesyl**, the C15 polymeric unit that occurs in **bacteriochlorophylls** *a* and *b*, two of the light-harvesting pigments present in primitive (non-oxygen-producing) photosynthetic bacteria. **Phytyl,** composed of four isoprene subunits, is the C20 polymer that occurs in **chlorophyll** *a*, the green-colored pigment used to carry out advanced oxygen-producing photosynthesis in cyanobacteria, algae, and higher plants. Not shown in Figure 4.2 are two even larger isoprene polymers, C30 squalene (the building block of cholesterol, composed of six isoprene units) and C40 carotene (the orange to red pigment made up of eight isoprenes that gives leaf-shedding trees their autumnal glory).

There is regularity and, thus, simplicity in these isoprene polymers, and cellulose is similar—a monotonous sequence of simple glucose monomers, repeated over and over again. This regularity and the resulting lack of variety in such polymers seem unexpected.

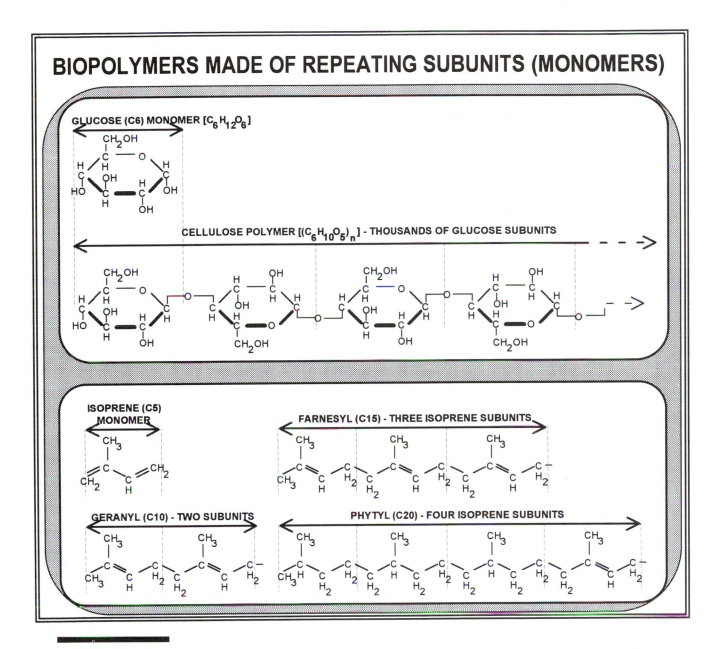

FIGURE 4.2

Biological polymers made of repeating subunits. Thousands of six-carbon glucose monomers (above) are linked together like beads on a string to form the very long cellulose polymer. Monomers of the five-carbon (C5) hydrocarbon isoprene (below) can be paired together to form the C10 polymer geranyl, or can be linked in linear groups of three (C15 farnesyl) or four (C20 phytyl).

In principle, the 20 carbon atoms and 39 hydrogen atoms that make up phytyl, the C20 isoprene polymer, could be combined to produce more than 300,000 different biochemicals, but only one of these, the phytyl polymer, is common in the living world.

Life is remarkably conservative—using only a minuscule portion of the almost limitless array of chemical combinations potentially available for polymer formation. The reason for this is that each of the many different types of biopolymers is constructed (that is, **biosynthesized**) step-by-step by a specific sequence of chemical reactions, **biosyn-**

thetic pathways that themselves include only a very small part of the huge number of ways that biochemical compounds might possibly be made. The journey to the final product of such pathways begins with a small step, usually the addition of atoms of CHON to a simple compound only a few carbons in length. In each new step, atoms or groups of atoms are added to produce molecules of increasing size. For cellulose, this is how the glucose monomer is initially synthesized; and for the isoprene polymers, this is the source of the isoprene building blocks. The series of steps leading to these monomers—like *all* steps in *all* biosynthetic pathways—are catalyzed, helped to occur, by **enzymes** (biochemicals that speed up chemical reactions), with each step in the molecule-making sequence set in motion by its own special enzyme. However, once the building blocks are made, the same type of enzyme can be used over and over to link monomers to the growing molecule. That is, the type of enzyme that can join together two isoprene units can also add an isoprene unit to the two-unit polymer, or to a polymer composed of three, or four, or virtually any other number of subunits. It is this repeated use of identical building blocks, linked together by reactions catalyzed by enzymes of the same type, that produces the orderly repeating pattern of polymers.

The way that a polymer is made, the pathway leading to its construction, is a legacy of evolution. Just like other parts of organisms (bony skeletons or teeth, for example), *biosynthetic pathways evolve over time.* Most pathways evolve merely by tacking on new steps at the ends of a preexisting sequence of reactions. For this reason, microbes living today that are directly descended from organisms of the very ancient past commonly contain only the early-evolving first several steps of such pathways, whereas advanced multicellular organisms contain much longer pathways that incorporate the early steps as well as many later additions and modifications.

Biosynthetic pathways provide a prime example of one of the most important take-home lessons in all of biology, namely, that *evolution is conservative and economical.* Rarely does evolution give rise to an absolutely new structure, an entirely new enzyme, a wholly novel way of life. Rather, the evolutionary process almost always modifies and builds on successful systems that already exist. This is understandable, perhaps even predictable. After all, major changes in virtually any well-tuned system are likely to take time, cost energy, and cause more harm than good. Simply put, there is no need for the evolution of new building blocks, new enzymes, new ways of life if minor modification of the ones at hand will do the job. The **Principle of Conservatism and Economy** is a hallmark of the evolutionary process, one that helps to solve such interesting puzzles as why animals breathe oxygen, why the ecosystem contains both the eaters and the eaten, and why plants are not intelligent.

The Necessities and Strategies of Life

At present, there are nearly 2 million species of living organisms known to science (and perhaps three to five times as many yet undiscovered). To keep alive, these different organisms need to satisfy just two basic necessities. Because all are made of CHON, all need a source of CHON. And in order to use that CHON, all need a source of energy. Thus, there are **two necessities of life—CHON and energy**. (To evolve, organisms also need to reproduce, but that, too, requires CHON for the developing offspring, and energy. Reproduction, and therefore evolution, can take place only if the two necessities are met.)

The key to living is to satisfy the two necessities. To meet these needs, there are only **two strategies of life—the plant-like strategy of autotrophy and the animal-like strategy of heterotrophy.**

Plants and plant-like organisms (single-celled algae and photosynthetic microbes, for example) are **autotrophs** (from the Greek, *autos*, self, and *trophos*, feeder; literally, "self-feeders"). Autotrophs satisfy their need for CHON by absorbing simple gases and nutrients (carbon dioxide, water, nitrate, and phosphate, for example) from their environment.

Most autotrophs are **photosynthetic** (and for this reason are referred to as **photoautotrophs**) organisms able to grow by using light energy to incorporate the absorbed CHON into molecules of glucose and other biological compounds. During this process, a portion of the harvested light energy is transferred to the chemical bonds that link together the atoms of C, H, O, and N, energy that can later be liberated and used by the organism. In essence, photoautotrophs are both "self-builders" *and* "self-eaters." They use CHON from the environment and the energy of light to build up biological compounds, and they later release energy from these compounds by breaking them down into smaller and smaller units. This process, the enzyme-aided buildup of CHON-containing compounds and their enzymatic breakdown to yield energy, is termed **metabolism**.

Animals and animal-like organisms (protozoans, fungi, and most nonphotosynthetic microbes) exhibit the other strategy, **heterotrophy** (from the Greek *heteros*, other, and *trophos*, feeder; literally "feeders on others"). Heterotrophs obtain their CHON from the food they eat; they obtain their energy by breaking down these foodstuffs and releasing the chemical energy stored in the bonds that link together the CHON atoms in that food. Thus—contrary to conventional wisdom—"simple" autotrophic plants are actually *more* complicated than heterotrophic animals. Autotrophs are able both to make food and to metabolize it to yield energy, but heterotrophs can carry out only the latter half of the process. That is why heterotrophs are the eaters and autotrophs are the eaten (and a major reason why humans depend on plant life for existence).

Thus, there are only two necessities of life—CHON and energy—and only two strategies—(photo)autotrophy and heterotrophy, but the evolutionary history of this system has additional lessons to reveal. Each of the basic metabolic strategies, photoautotrophy and heterotrophy, occurs in two versions, *a primitive (early-evolved) and an advanced (later-evolved) form.* As shown in Figure 4.3, the difference between the two forms depends on

METABOLISM	"CHON" SOURCE	ENERGY SOURCE
AEROBIC HETEROTROPHY Oxygen-requiring bacteria, fungi, protists, animals	**FOODSTUFFS PRODUCED BY AUTOTROPHS**	**AEROBIC RESPIRATION** $O_2 + GLUCOSE \longrightarrow 36\ ATP$ (bond energy of consumed foodstuffs)
AEROBIC PHOTOAUTOTROPHY Oxygen-consuming and -producing cyanobacteria, protists, plants	**OXYGENIC PHOTOSYNTHESIS** $CO_2 + H_2O \longrightarrow GLUCOSE + O_2$ (plus N_2 - fixation in cyanobacteria)	**AEROBIC RESPIRATION** $O_2 + GLUCOSE \longrightarrow 36\ ATP$ (bond energy of photosynthesized glucose)
ANAEROBIC PHOTOAUTOTROPHY *Non*-oxygen-requiring photosynthetic bacteria	**ANOXYGENIC PHOTOSYNTHESIS** $CO_2 + H_2S \longrightarrow GLUCOSE$ (plus N_2 - fixation)	**GLYCOLYSIS** $GLUCOSE \longrightarrow 2\ ATP$ (bond energy of photosynthesized glucose)
ANAEROBIC HETEROTROPHY *Non*-oxygen-requiring fermenting bacteria	**FOODSTUFFS PRODUCED BY AUTOTROPHS** (or formed abiotically in early environment; plus N_2 - fixation)	**GLYCOLYSIS** $GLUCOSE \longrightarrow 2\ ATP$ (bond energy of consumed foodstuffs)

FIGURE 4.3

The necessities of life, sources of CHON and energy, are met by two metabolic strategies, heterotrophy and autotrophy. Each of these occurs in a primitive (anaerobic) and a later-evolved (aerobic) form.

the presence of molecular oxygen (O_2). The type of photosynthesis carried out by primitive photoautotrophs does *not* generate oxygen as a by-product (and is therefore an **anoxygenic** process), and neither it, nor primitive heterotrophy, requires the presence of oxygen (that is, both are **anaerobic**). In contrast, the advanced, later-evolved versions of both strategies occur only in the presence of oxygen—advanced photosynthesis is oxygen-generating (**oxygenic**) and all organisms that have later-evolved forms of autotrophy and heterotrophy are oxygen-using **aerobes**, organisms that depend on oxygen to breathe (that is, whether their metabolism is plant-like or animal-like, they all carry out **aerobic respiration**).

Two necessities; two strategies; and two forms of each strategy, one primitive (anaerobic) and one advanced (aerobic)—these simple observations go a long way toward explaining the fundamental nature of the total ecosystem, because exactly the same rules apply to all biological levels, from microbes to man. Once this ecologic structure became established by primitive microbes—long before the appearance of plants or animals of any type—it remained essentially unchanged and has been carried over to all later-appearing ecosystems, even to that of the present day.

The Evolutionary Tree of Life

How could this be true? Microbes obviously differ from humans. Does it make sense to imagine that microbial communities of billions of years ago could have any relevance to the complex global ecosystem of today? Moreover, such ancient microbes could only be remotely related to present-day plants and animals—how could their metabolic inventions have been passed along generation to generation over the countless intervening millennia?

To answer these questions, let us look at the **Tree of Life** shown in Figure 4.4 (and discussed also in Chapters 1 and 3), a diagram that summarizes evolutionary relations among the major biological groups (or **lineages**) that inhabit the present world. **Phylogenetic** trees of this sort are based on comparisons of the chemical makeup of a particular type of **rRNA** (ribosomal ribonucleic acid), a form known technically as 16S rRNA, isolated from various living organisms. Ribosomal ribonucleic acids are among the best of all biological compounds for building such a tree because they occur in **ribosomes** (the cellular machinery used to make proteins) and are therefore present in *all* forms of life. The diagram shows that there are three main branches of the Tree of Life, microbes of the **Eubacteria** and **Archaebacteria** and advanced "higher" organisms, the **Eukaryotes**. As expected, eukaryotic plants and animals (at the upper right in Figure 4.4) are distantly related to the primitive earlier-evolving microbes. How could metabolic inventions be passed along unchanged over such a vast evolutionary distance?

The solution to this puzzle is shown in Figure 4.5. Two particular groups of eubacterial microbes, *cyanobacteria* and *purple bacteria*, are central to the story. The advanced form of photoautotrophy, oxygenic photosynthesis, was first invented by cyanobacteria, evidently as early as 3.5 billion years ago when Earth was in its infancy, only one-quarter of its current age. Very much later in geologic time, perhaps as recently as 2 billion years ago, descendants of cyanobacteria were swallowed up by later-evolving, single-celled eukaryotes. Because the engulfed (but undigested) cyanobacteria carried with them all of their metabolic machinery, a mutually beneficial **symbiotic** relationship became established, one in which the cyanobacteria, now living within the eukaryotic host, produced food for the host by oxygenic photosynthesis and the host cell provided protection for its newly acquired internal food factory. As this **endosymbiosis** continued to develop, the engulfed cyanobacteria lost their ability to live outside the host cell and evolved within it to become **chloroplasts**, the photosynthethic **organelles** (small membrane-enclosed bodies) that are present in all photoautotrophic eukaryotes living today. As a result of a sim-

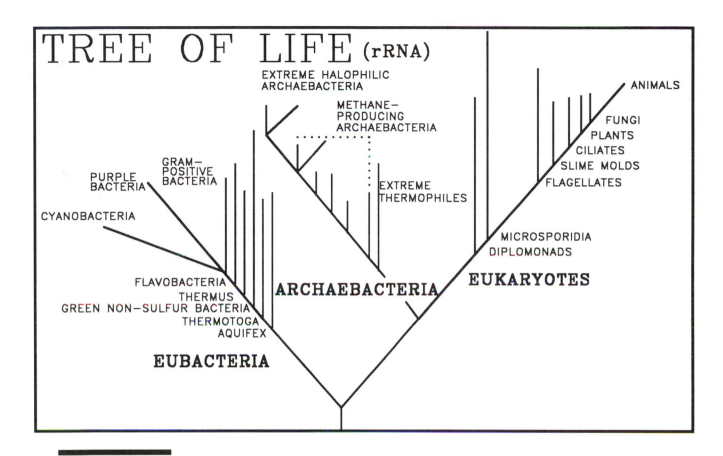

FIGURE 4.4

Phylogenetic Tree of Life, based on molecules of "16S" rRNA (one of several types of RNA occurring in protein-manufacturing ribosomes), the tree shows the ancestor-descendant relations among major types of eubacterial, archaebacterial, and eukaryotic modern organisms. The Archaebacteria may actually include two major lines of descent (Rivera and Lake, 1992), rather than only one as is shown here. (Based on data from Woese, 1987; Iwabe et al., 1989; and Woese et al., 1990.)

ilar evolutionary endosymbiosis, originally free-living, oxygen-breathing purple bacteria that had been swallowed up by eukaryotic hosts evolved to become **mitochondria**, the organelles in which aerobic respiration takes place in all oxygen-breathing eukaryotes.

Despite the vast evolutionary distance that separates plants and cyanobacteria, it is therefore not surprising that they share the same photosynthetic machinery; plant chloroplasts are simply modified cyanobacteria. The same is true of the way animals, plants, and eubacteria breathe. All groups share the same aerobic metabolism because animal and plant mitochondria are merely evolutionarily modified eubacteria. The evolutionary process has simply altered and built on successful systems that previously existed. The take-home lesson continues to hold: *Evolution really is both conservative and economical.*

Nevertheless, questions remain. Although the process of oxygenic photosynthesis may be identical in higher plants and cyanobacteria, how did it *initially* come into being? Is it a borrowed and modified version of some earlier metabolic invention, or did it arise from scratch in the earliest cyanobacterium billions of years ago? Similarly, what are the roots of aerobic respiration? The ability to breathe requires the presence of molecular oxygen, but how could such a process get started if, as most scientists think, there was no oxygen

FIGURE 4.5

Organelles of eukaryotic cells are evolutionary derivatives of endosymbiotic, but originally free-living eubacteria. Endosymbiotic cyanobacteria evolved within eukaryotic hosts to become the chloroplasts of algae and higher plants. Similarly, the mitochondria of oxygen-breathing aerobic eukaryotes evolved by modification of endosymbiotic purple bacteria.

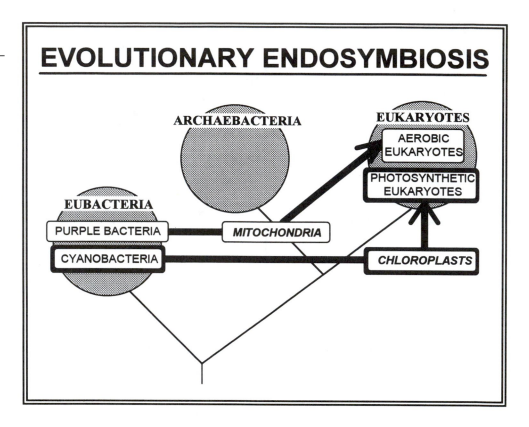

in the early atmosphere? In short, if evolution can only build on something that already exists, how could a totally new biological invention ever originate?

THE EARLIEST STRATEGY OF LIFE: ANAEROBIC HETEROTROPHY

Glycolysis: Energy from Sugar Fermentation

Among the most successful early forms of life were those that could carry out **glycolysis** (Figure 4.6, left column), a 10-step metabolic process during which the six-carbon sugar glucose is split in half to produce smaller (C3) molecules of **pyruvate** and a small amount of energy is released (liberated by breaking apart the chemical bonds in glucose). As in most such energy-generating processes, a portion of the energy released is transferred to molecules of **ATP** (adenosine triphosphate), two ATPs being formed during breakdown of each molecule of glucose sugar.

Why is it "certain" that glycolysis was such an early evolutionary invention? The best evidence comes from the fact that this metabolic process is a fundamental feature of *all* living systems. Glycolysis is universal, and because a many-step reaction sequence of this sort is unlikely to have originated more than once in the history of life, it must have been present during life's earliest stages. Moreover, glycolysis is a type of **fermentation** (that is, anaerobic metabolism); it therefore does not require molecular oxygen, and that fits with geologic evidence indicating that the early environment was essentially **anoxic** (devoid of O_2). Glycolysis is simple—unlike most later-evolving metabolic processes, it takes place entirely in the watery cytosol of cells (that is, it does not require membranes or organelles); chemically, it is the simplest energy-generating process known in living systems; and it yields relatively little energy (only 2% of that potentially available from the breakdown of glucose), much less than that provided by the advanced energy-yielding process of aerobic respiration.

Laboratory experiments show that limited quantities of sugars, the fuel for glycolysis, would have been formed by chemical reactions occurring in the primordial anoxic atmosphere and oceans (manufactured by **abiotic syntheses**, chemical processes independent of life). However, these experiments also show that in addition to glucose, many

FIGURE 4.6

Metabolic pathways of glycolysis, glucose biosynthesis, and the dark reactions of oxygenic photosynthesis are fundamentally similar. The first to evolve was probably glycolysis (known also as sugar fermentation; left column). Reversal of this pathway and evolution of a few new enzymes resulted in development of the pathway for glucose biosynthesis (center column). Almost all the pathway for glucose biosynthesis was carried over intact to form the sequence of glucose-synthesizing dark (non-light-requiring) reactions that occur in cyanobacteria, algae, and higher plants during oxygen-producing photosynthesis (right column).

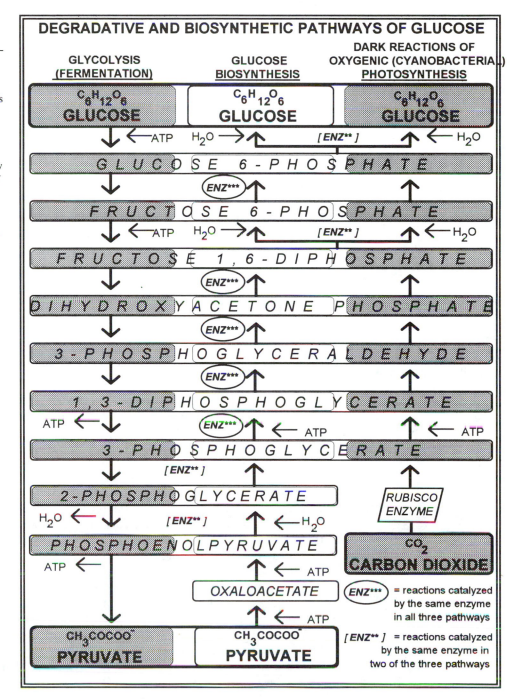

other sugars would have been made. Why is glucose the universal fuel of life? Possibly because glucose is an especially robust molecule. Of the 16 different six-carbon sugars that can be made, glucose is apparently the least susceptible to breakdown by environmental changes in temperature, acidity, and the like. In the harsh primordial environment, glucose may have been the most likely sugar to have accumulated, and may thus have been particularly available as a foodstuff for early life.

Thus, nonbiologically formed glucose may have served as a food source for early microbes. Such microorganisms would therefore have been *heterotrophs* and, because of the absence of molecular oxygen, necessarily *anaerobic*. However, as these microbes multiplied, they would have rather rapidly begun to exhaust the limited glucose fuel supply; without further evolution they would have starved to the edge of extinction, literally eating themselves out of house and home. What happened next? Certainly, something did happen; otherwise, glycolysis would have been lost forever, rather than having been carried over into each individual organism (including you and me) that has ever since inhabited our planet.

Glucose Biosynthesis: A New Source of Fuel

Because not enough glucose was being produced nonbiologically in the early environment, the solution for primitive anaerobic heterotrophs was to manufacture it themselves by adding a pathway for glucose biosynthesis. The foundation for just such a pathway was readily at hand—the glycolysis pathway—but it had to operate in the opposite direction. As shown in Figure 4.6, glucose biosynthesis involves 11 enzyme-mediated steps, of which 7 use exactly the same enzymes as those of glycolysis, but perform their chemical reactions in the opposite direction. The **genes** for the glycolysis enzymes (that is, the biochemical blueprints for their construction) are housed in **DNA** (deoxyribonucleic acid), the hereditary molecule of life. To evolve the new pathway for glucose biosynthesis, an additional set of enzymes was needed. However, rather than construct an all-new set of genes and enzymes, evolution was conservative and economical. Without changing the successful glycolysis system, 7 of the enzymes already in use for glycolysis were "recruited" (that is, they were borrowed for use in the new glucose biosynthetic system, probably by duplication of their genes) and a few new enzymes were added.

How can the same enzyme help a chemical reaction to go both forward *and* backward, and what determined which enzymes should be changed? Imagine that a train is moved along a track. If the railroad track is level, the same amount of energy is required to push the train forward and to pull it backward. This is also true of most chemical reactions set in motion by enzymes; with essentially equal ease, the same enzyme can help push the reaction in one direction or pull it in the opposite direction (that is, the reactions are **reversible**). But this is not true if the track is on a steep slope. To move a train uphill, extra energy must be added; but as a train moves downhill, energy is released because of gravity. Similarly, in enzyme-aided pathways, reactions that either require or release extra energy (provided by or produced in the form of ATP) usually can be carried out easily in only one direction (and are therefore essentially **irreversible**). In glycolysis, both the first and third steps require input of ATP (Figure 4.6, upper portion of left column), and the final steps in the sequence are energy-producing (Figure 4.6, bottom of left column); these are the three irreversible steps in which new enzymes were needed for glucose biosynthesis.

Thus, the shortage of glucose was evidently offset by the advent of glucose biosynthesis, a new metabolic process that evolved simply by reversal and modification of the preexisting pathway of glycolysis. The new system was built on the foundation provided by a pathway that had already existed. However, this solution to the fuel shortage would have served only as a stopgap measure, because three times as much energy (in the form of six ATPs) is needed to biosynthesize glucose by this system than is produced by glucose fermentation. An organism cannot long survive if it uses more energy (to synthesize glucose, for example) than it generates (in this case, to break down glucose). In biology, as in a bank account, deficit spending is never more than a relatively short-term solution—ultimately, outgo must be balanced by income. The need for energy balance was soon answered by the evolution of a far more cost-effective way to manufacture glucose, a new system driven by the power of light. But before that system could evolve, yet another problem had to be solved.

What Was the Source of the N in CHON?

Glucose sugar ($C_6H_{12}O_6$) is composed of carbon, hydrogen, and oxygen. However, proteins (including most enzymes), nucleic acids (genetic material), and ATP (energy-storage molecules) contain nitrogen (N) in addition to CHO (see Figure 4.1). Anaerobic heterotrophs thrive if they are provided two basic nutrients—glucose (for CHO and energy) and ammonia (the source of N). Initially, this may also have been true on early Earth, but ammonia, NH_3, would soon have been in short supply. The chemical bonds that tie together the nitrogen and hydrogen atoms in ammonia are easily broken by high-energy ultraviolet (UV) light. Thus, because there was essentially no molecular oxygen (O_2) in the early environment and a UV-absorbing atmospheric ozone (O_3) shield did not exist, ammonia would have been rapidly destroyed. In fact, a quantity of ammonia equivalent to *all* the enormous amount of molecular nitrogen (N_2) in the present-day atmosphere would have disappeared in less than 100,000 years, a mere instant in geologic time. The availability of nitrate, NO_3^-, the other source of nitrogen used by many organisms, would also have been limited. At present, large quantities of nitrate are formed when gaseous oxygen and nitrogen react during lightning storms, reactions that could not have occurred in the early oxygen-deficient atmosphere.

The need for usable nitrogen coupled with the scarcity of ammonia and nitrate presented a major problem for life. Only one feasible source of nitrogen was left—the molecular nitrogen of the atmosphere. However, the atoms in N_2 are tenaciously locked together; a specialized enzyme system was therefore needed to break them apart. The driving force of this system is **ferredoxin**, a particularly interesting compound because it is used not only in the **fixation of nitrogen** (the bonding of hydrogen to nitrogen to form ammonia) but also in many other biochemical processes. The new ferredoxin-driven nitrogenase enzyme complex (the *Nif* **complex**) originated in response to a critical shortage in the availablity of usable nitrogen. Evolution usually is energetically cost-effective; almost always, newly evolved mechanisms require input of less energy (or provide more energy to the organism) than previously established systems. However, N_2-fixation is energetically *very* costly, far more so than use of ammonia or nitrate. In fact, because of the large amount of cellular energy required to break apart molecular nitrogen, N_2-fixation is used as a last resort by living systems, employed only when other nitrogen supplies have already been exhausted.

Numerous lines of evidence indicate that the ferredoxin-driven *Nif* complex evolved quite early. For example, although N_2-fixation is widespread among **prokaryotes** (microbes of the Eubacteria and Archaebacteria), it is entirely absent from later-evolving eukaryotes (Figure 4.7). Moreover, like many other metabolic processes that originated in the early oxygen-deficient environment, operation of the *Nif* complex is brought to a standstill by the presence of trace amounts of molecular oxygen. Nitrogen-fixation is therefore common among anaerobic forms of life (microbes ranging from heterotrophs to anoxygenic photoautotrophs, sulfate-reducers, and methane-producing archaebacteria), and although the process can also be carried out by some *aerobic* cyanobacteria, advanced mem-

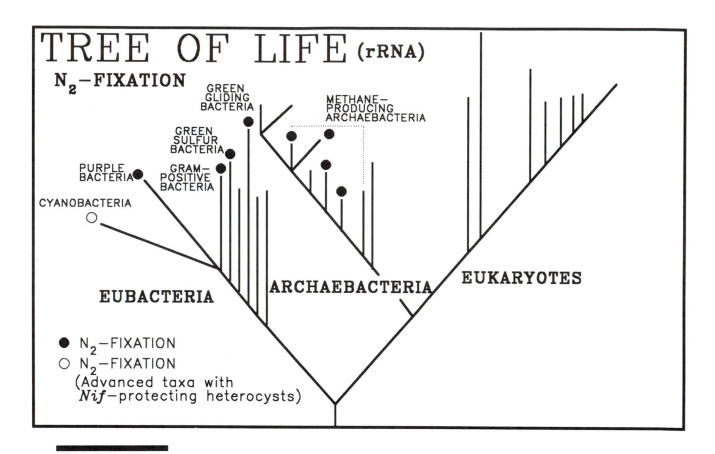

FIGURE 4.7

The metabolic ability to fix atmospheric nitrogen (N₂) is a primitive, early-evolved trait, characteristic of numerous types of eubacterial and archaebacterial prokaryotic microbes. (See Figure 4.4 for sources of data.)

bers of this group have special thick-walled cells that protect the *Nif* complex from contact with oxygen (see also chapter 1).

Thus, ferredoxin and the *Nif* complex may have played a central role in solving the nitrogen crisis faced by primitive anaerobic heterotrophs. Where did the ferredoxin come from?

Ferredoxin is a **protein**, a polymer composed of **amino acids** (small compounds that contain -NH₂, the so-called amino group, and -COOH, the carboxylic acid group) linked together in a simple beads-on-a-string array. For example, the ferredoxin shown in Figure 4.8, isolated from a modern microbial anaerobic heterotroph, *Clostridium pasteurianum*, is composed of a linear sequence of 55 amino acids (each denoted in Figure 4.8 by a three-letter code, with Ala representing the amino acid alanine; Asp, aspartic acid; Ser, serine; Gly, glycine; and so forth, as discussed in Chapter 2). Careful studies of this ferredoxin (Eck and Dayhoff, 1966) revealed much about its origin and evolutionary history. In principle, each of the 55 positions filled by amino acids in this molecule could be occupied by any of the 20 amino acids that occur in proteins. However, comparison of the two halves of the molecule shows that this is not the case. In fact, there are 12 different sites at which identical amino acids occur at the same position in each half of the molecule (the boxed amino acids shown in the upper panel of Figure 4.8), far too great a coincidence to be a result of chance. It seems certain, therefore, that this ferredoxin was derived by the doubling of an earlier-evolved proto-ferredoxin that contained only half as many amino acids (including all 12 of those occurring at the same site in the two halves of the molecule). Further analyses show that an unexpectedly large number of amino acids occur in cycles of four; indeed, 13 pairs of amino acids occur at every fourth position in the molecule (separated by three other amino acids) or in some multiple of four (Figure 4.8, middle panel). Of all the amino acids, Ala, Asp, Ser, and Gly stand out as being especially common at every fourth position (Figure 4.8, lower panel).

These clues have been used to reconstruct the origin and evolution of the molecule (Figure 4.9). Early in the history of life, a primitive anaerobic heterotroph may have contained a precursor molecule composed of the amino acid quartet Ala-Asp-Ser-Gly (1-2-3-4, in Figure 4.9). If so, the gene for this quartet must have been duplicated several times, giving rise to a much longer gene that contained the blueprint for construction of a proto-ferredoxin molecule made up of 28 amino acids, a repeating sequence of seven of the original quartets. An additional amino acid was then evidently inserted in the center of

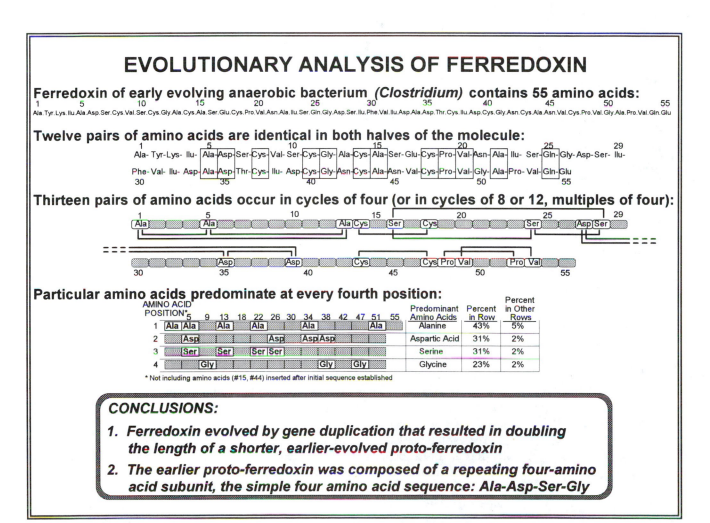

FIGURE 4.8

Microbial ferredoxin evolved by gene duplication, probably from a proto-ferredoxin composed of a repeating quartet of four amino acids.

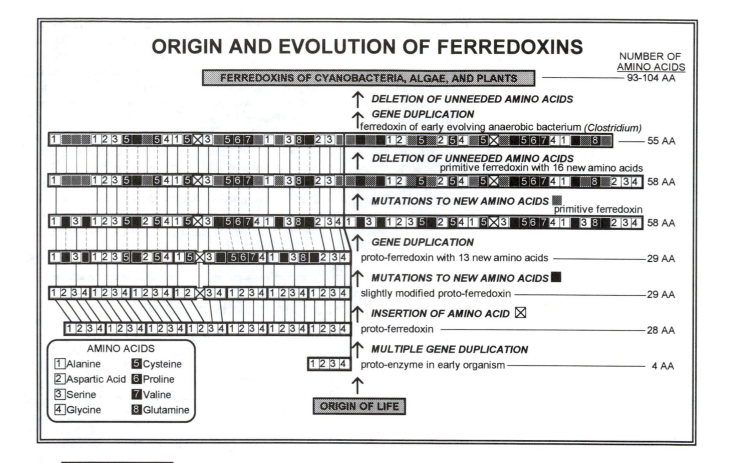

FIGURE 4.9

Origin and evolution of ferredoxins.
(Based on the studies of Eck and
Dayhoff, 1966.)

the molecule and, as a result of **mutations** (changes in the genetic blueprint), certain of
the original amino acids were replaced (for example, by amino acids 5, 6, 7, and 8 in Figure 4.9). The gene for the resulting 29 amino acid polymer was duplicated to produce a
primitive ferredoxin composed of 58 amino acids; additional mutations occurred; and three
unnecessary amino acids were deleted from one end of the molecule to produce the 55
amino acid ferredoxin of the modern-day microbe.

The evolutionary history of microbial ferredoxin is thus rather simple, chiefly involving repeated gene duplication, augmented by occasional mutations and infrequent amino
acid insertions or deletions. **Gene duplication** (doubling of the protein-producing genetic
blueprint), the principal mechanism driving the evolution of this molecule, is widespread
among microbes, and may have been especially common during life's early development
when sources of CHON and energy were limited. Studies of modern microbes show that
a few to many genes are duplicated in about one of every 1,000 individuals. This means
that in a population of a billion microbes (a normal community for such minute forms of
life) there are about a million mutant individuals that contain duplicated genes. Plentiful
grist for the evolutionary mill! Moreover, laboratory studies of microbes subjected to extreme starvation—conditions probably not unlike those of the primitive Earth—show that
survival occurs almost exclusively (up to 96% in one experiment) in those mutants in

which genes have been duplicated to produce extra copies of metabolically crucial enzymes.

The evolutionary history of ferredoxin, however, does not stop with microbes. Ferredoxins also occur in eukaryotes (for functions other than nitrogen fixation), and these later-evolving ferredoxins tend to be about twice as long as the earlier-evolved forms of the molecule. Thus, advanced ferredoxins also evolved by gene duplication. None of the numerous ferredoxins, whether in prokaryotes or eukaryotes, is absolutely *new*. Except for the initial quartet of amino acids, all ferredoxins are built on a foundation that was already well established. Significantly, however, the early evolutionary use of ferredoxin was not cost-effective; use of N_2 as a nitrogen source required expenditure of decidedly more energy, not less, than systems established earlier. To have become so widespread among early-evolving prokaryotes, ferredoxin-driven N_2-fixation must have been central to the survival of life.

■

AIR AND LIGHT: A NEW SOURCE OF GLUCOSE

Making Glucose by Photosynthesis

The earliest strategy of life was anaerobic heterotrophy: CHO was provided by glucose, N by fixation of atmospheric nitrogen and use of the limited quantities of ammonia and nitrate available in the environment, and energy by glucose fermentation (glycolysis). However, a new way was needed to manufacture the glucose sugar needed to fuel the system in copious amounts (unlike abiotic syntheses) and that was energetically cost-effective (unlike glucose biosynthesis). This need was met by harnessing the power of light in the process of photoautotrophy.

As shown in Figure 4.10, most eubacterial lineages contain members that are photosynthetic, the most primitive of which carry out the anaerobic process of bacteriochlorophyll-based anoxygenic photoautotrophy. How did this process arise? An early development must have been the origin of the biosynthetic pathway leading to bacteriochlorophylls, the light-harvesting pigments on which all types of anoxygenic photosynthesis are based. However, in this pathway (Figure 4.11), biosynthesis of light-harvesting pigments is closely tied to formation of chlorophyll *a*, the photosynthetic pigment of oxygen-producing cyanobacteria. The chlorophyll-type pigments of *all* photoautotrophs are similar, regardless of whether oxygen is produced as a photosynthetic by-product, because all are manufactured by similar means; thus, all have very similar chemical structures (Figure 4.12).

Were chlorophyll-type molecules first used for photoautotrophy, or were they initially used for something else? As Figure 4.13 shows, *all* (bacterio)chlorophyll-based prokaryotic **photosystems** are used not only for photoautotrophy, but also for **photoheterotrophy**—light energy captured by these pigment systems is used to speed uptake into cells of CHON-containing compounds from the environment; once assimilated, the compounds are broken down by heterotrophic processes. Because light-driven uptake of organic compounds (known technically as **photoassimilation**) is widespread, prokaryotic photosystems were probably first used as a means to improve efficiency of earlier established anaerobic heterotrophy, not for autotrophy. This is *another example of the conservatism and economy of evolution.*

As discussed earlier, there are two types of prokaryotic photosynthesis, one primitive and anoxygenic (non-oxygen-producing) and the other advanced and oxygen-producing (Figure 4.14). How do the two processes differ? The most obvious difference is the source of the hydrogen that is combined with carbon dioxide to form glucose. In primitive anoxygenic photosynthesis, hydrogen is usually supplied either in the form of molecular hy-

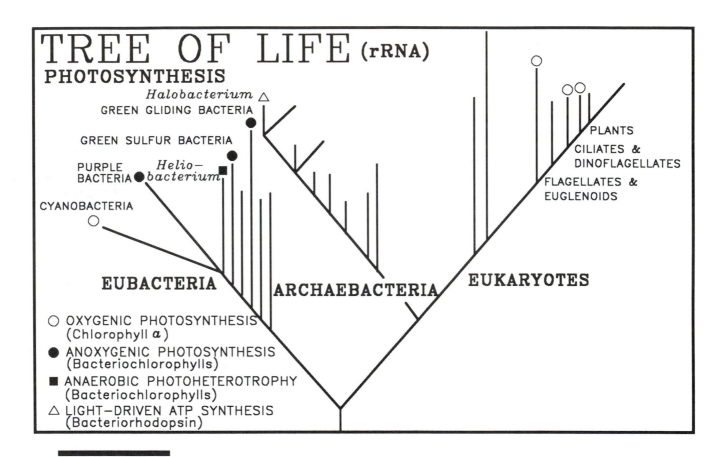

FIGURE 4.10

Photosynthesis is a primitive trait, characteristic of many lineages of early-evolving eubacteria. (See Figure 4.4 for sources of data.)

drogen, H_2, or as hydrogen sulfide, H_2S. In contrast, in advanced oxygenic photosynthesis, hydrogen is provided by water, H_2O; hydrogen atoms are split away from the water molecule leaving oxygen as a photosynthetic by-product. The sources differ in the ease with which hydrogen atoms can be extracted from them (see Figure 4.13). Use of H_2 as a hydrogen source (the **electron donor**; Figure 4.13) does not require much energy, and separation of hydrogen from H_2S requires only somewhat more (78 kilocalories). However, the hydrogen and oxygen atoms in water are tightly linked, and 50% more energy (a total of 118 kilocalories) must be spent to split them apart. Thus, the earlier-evolved anoxygenic process requires less energy, and advanced oxygenic photosynthesis demands more.

Therefore, by harnessing the power of light, the development of photoautotrophy provided a valuable new means to manufacture glucose, but the way this process evolved raises interesting questions. If evolution is usually energetically cost-effective, why does the advanced process require *more* energy than that needed by the system from which it was evolved? Doesn't this fly in the face of the notion that evolution is economical?

Anoxygenic and Oxygenic Photoautotrophy

Oxygen-producing photosynthesis is more complicated than the earlier evolved anoxygenic version. As shown in Figure 4.15, the process is composed of two separate light-

FIGURE 4.11

Bacteriochlorophylls and chlorophylls are manufactured by virtually the same biosynthetic pathway. Each of the boxes in the pathways represents one or a few enzyme-driven biochemical reactions.

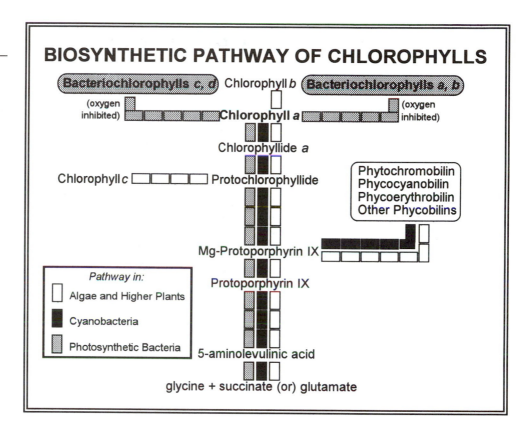

BIOSYNTHETIC PATHWAY OF CHLOROPHYLLS

Bacteriochlorophylls c, d   Chlorophyll b   Bacteriochlorophylls a, b

(oxygen inhibited)   **Chlorophyll a**   (oxygen inhibited)

Chlorophyllide a

Chlorophyll c   Protochlorophyllide

Phytochromobilin
Phycocyanobilin
Phycoerythrobilin
Other Phycobilins

Mg-Protoporphyrin IX

Protoporphyrin IX

Pathway in:
☐ Algae and Higher Plants
■ Cyanobacteria
▨ Photosynthetic Bacteria

5-aminolevulinic acid

glycine + succinate (or) glutamate

dependent events and associated biochemistries, Photosystems I and II that together make up **light reactions** and another series of chemical reactions that do not require light (**dark reactions**). With the help of a manganese-containing enzyme, water is split by light energy captured by chlorophyll *a* to produce hydrogen and molecular oxygen. At the same

FIGURE 4.12

Chlorophyll *a*, the light-harvesting pigment of cyanobacterial oxygen-producing photosynthesis, is composed of four nitrogen-containing rings (pyrroles) to which are attached various small chemical groups. This same basic structure (shown by thick lines) also occurs in the bacteriochlorophylls used in non-oxygen-producing photosynthesis.

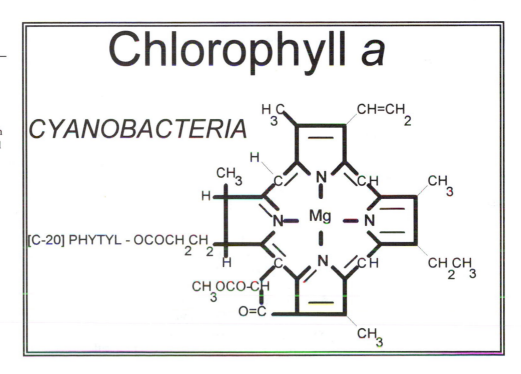

Chlorophyll *a*

CYANOBACTERIA

[C-20] PHYTYL - OCOCH$_2$CH$_2$

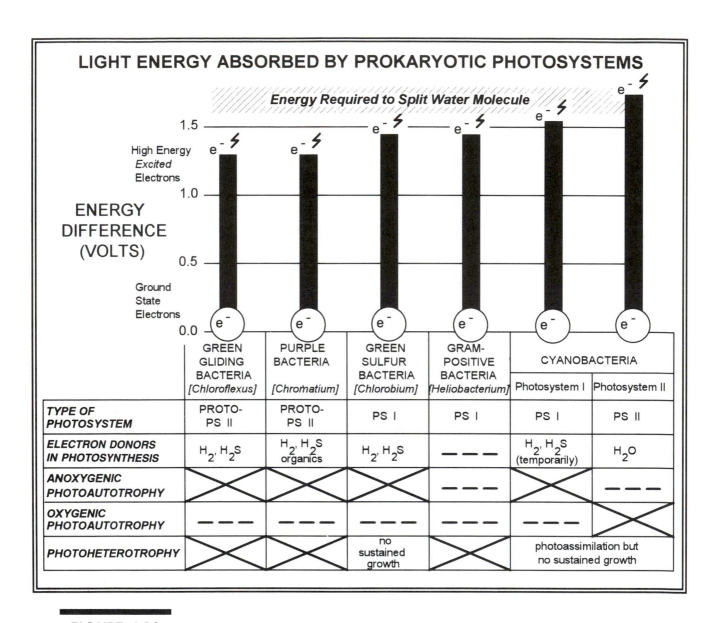

FIGURE 4.13

More energy is required to split the water molecule (in cyanobacterial Photosystem II) than to extract hydrogen from the other electron donors used in prokaryotic photosystems. In the table at the bottom of the figure, *X* indicates that the organism is capable of carrying out a given process whereas a dashed line indicates the lack of such ability. Because all (bacterio)chlorophyll-containing prokaryotes can use light energy to speed uptake of organic compounds from their surroundings, it is likely that their pigment systems were first used in photoheterotrophy rather than in autotrophy.

time, in the first photochemical event of **Photosystem II**, light energy is transferred to electrons in atoms of chlorophyll *a*, causing the atoms to be energized to an **excited state** (a state in which the electrons contain much more energy than they do in their normal **ground state**). However, this excited state is not stable; to return to their original ground state, the energized electrons must give up the extra energy they have absorbed. The extra energy is released as the electrons are passed along a series of **electron carriers**, and a portion of this energy is fed into molecules of ATP. At the end of the transport system, the electrons are delivered to another chlorophyll a-containing site where a second photochemical event begins, **Photosystem I**. Light energy excites electrons in the chlorophyll at this site, and the energized electrons are passed to a large electron-accepting molecule (NADP), the last step in the light-requiring portion of oxygenic photosynthesis. Finally, the electrons stored in NADP are passed to molecules of CO_2 that have been taken in from the environment, and glucose, the end product of photosynthesis, is manufactured by a series of enzyme-aided dark reactions (see Figure 4.15).

The glucose-producing dark reactions of oxygenic (cyanobacterial) and anoxygenic (bacterial) photosynthesis are similar to each other, and both strongly resemble the glucose biosynthetic pathway that evolved earlier in anaerobic heterotrophs. In fact, the three processes differ basically only in the source of energy needed to manufacture the starting compound shared by all three systems, **3-phosphoglycerate** (also known as 3-phosphoglyceric acid, **3PGA**). In heterotrophs (see Figure 4.6, center column), cellular energy in the form of ATP is used to make 3PGA, whereas both in oxygenic cyanobacteria (see Figure 4.6, right column) and in anoxygenic bacterial photosynthesizers (Figure 4.16, left column) the necessary energy is provided by light.

Oxygenic and anoxygenic photosynthesis have much in common. Both processes use the same carbon source, CO_2, and have similar light-harvesting pigments, electron- and energy-generating light reactions, and glucose-producing dark reactions. Their descendant-ancestor relationship is well documented by the rRNA phylogenetic tree (see Figure 4.10). Moreover, numerous species of cyanobacteria are able to carry out *both* processes, switching from advanced oxygenic photosynthesis to the more primitive and less energy-consuming anoxygenic process when H_2S is locally available, a co-occurrence of the two systems that was probably also exhibited by the earliest-evolving oxygen-producers.

Although closely related, the two processes differ in important ways. In later-evolving, oxygen-producing photosynthesis, chlorophyll a is the light-capturing pigment; water serves as the source of hydrogen, and O_2 is a by-product; and two separate photosystems (I and II) are linked by a series of electron carriers. In contrast, in the earlier-evolved anoxygenic process, bacteriochlorophyll is the light-capturing pigment; water is *not* used as the source of hydrogen and O_2 is *not* produced as a by-product; and there is only a sin-

---

FIGURE 4.14

Two types of photosynthesis occur in prokaryotes: advanced, later-evolved, oxygen-producing photosynthesis and primitive, earlier evolved, *non*-oxygen-producing photosynthesis.

TYPES OF PROKARYOTIC PHOTOSYNTHESIS

Advanced Oxygenic (Oxygen-Producing) Photosynthesis:

CYANOBACTERIA ONLY

$$6\ H_2O\ +\ 6\ CO_2\ \longrightarrow\ C_6H_{12}O_6\ +\ 6\ O_2$$

water   carbon dioxide   glucose sugar   molecular oxygen

Primitive Anoxygenic (*Non*-Oxygen-Producing) Photosynthesis:

ALL PHOTOSYNTHETIC PROKARYOTES

$$12\ H_2S\ +\ 6\ CO_2\ \longrightarrow\ C_6H_{12}O_6\ +\ 12\ S\ +\ 6\ H_2O$$

hydrogen sulfide   carbon dioxide   glucose sugar   sulfur   water

$$12\ H_2\ +\ 6\ CO_2\ \longrightarrow\ C_6H_{12}O_6\ +\ 6\ H_2O$$

molecular hydrogen   carbon dioxide   glucose sugar   water

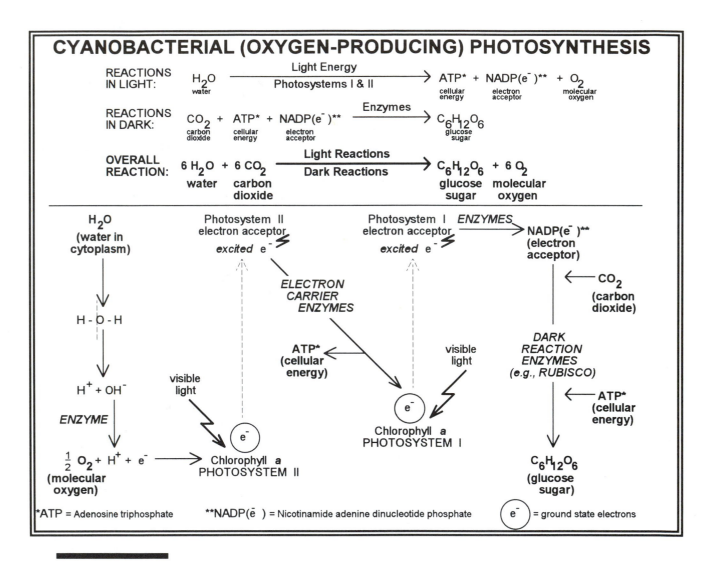

FIGURE 4.15

To generate glucose, oxygen-producing (cyanobacterial) photosynthesis involves a two-part sequence of light-requiring reactions followed by a series of enzyme-aided dark reactions.

gle photosystem with its associated electron carriers (similar either to Photosystem I, as in green sulfur bacteria, or to Photosystem II, as in green gliding and purple bacteria). Because oxygen-producing photosynthesis involves two, rather than only one photosystem, the later-evolved process is the more complicated of the two photoautotrophic processes, and it also requires more energy. These shortcomings are weighty; if they had not been offset by significant advantages, it is unlikely that the advanced process—a process that occurs not only in cyanobacteria but also in all present-day plants—would ever have become so successful.

What were these advantages? Part of the answer may stem from the availability of hydrogen on the primitive Earth. Although hydrogen sources used by anaerobic photosynthesizers were locally plentiful—hydrogen sulfide in hot springs and fumarolic volcanic vents, for example—they were not universally abundant. Water, in contrast, was both plentiful and ubiquitous. Use of water as a hydrogen source by oxygen-producing photoau-

FIGURE 4.16

The dark reactions of anaerobic bacterial photosynthesis are fundamentally similar to the citric acid cycle of aerobic respiration. The first to evolve was the cyclic pyruvate-synthesizing pathway of anoxygenic bacterial photoautotrophs (left column). Reversal of this pathway and substitution of a few new enzymes resulted in development of the citric acid cycle that occurs in all oxygen-breathing organisms (right column).

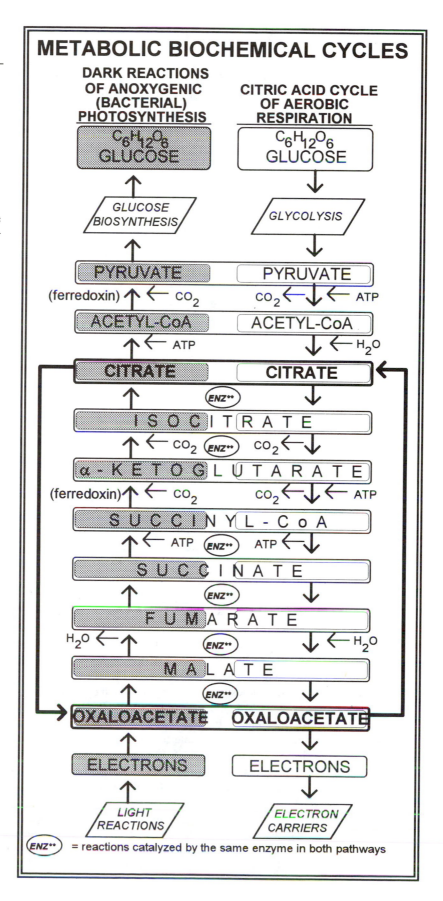

totrophs may therefore have enabled photosynthesizers to spread to a profusion of previously uninhabited locales.

Probably even more important, however, were the fundamental differences in oxygen relations of the two types of photosynthesis. Like N_2-fixation, anoxygenic photosynthesis is an anaerobic process; and, like the oxygen-sensitive *Nif* complex, enzymes required for the manufacture of bacteriochlorophylls are turned off by trace amounts of free oxygen (see Figure 4.11). In contrast, oxygen-producing photosynthesizers are aerobes, organisms that thrive in the presence of molecular oxygen. These differences in oxygen relations had enormous impact on the subsequent development of life.

The First Oxygen-Producers

Let us imagine what happened when the first oxygen-producers appeared. The microbes of this new mutant strain, the first true cyanobacteria, shared a shallow-water environment with the anoxygenic photosynthesizers from which they were derived. Like their parent stock, they were photoautotrophs; because both they and the parental strain depended on light, there was fierce competition for photosynthetic space. However, the mutants brought a telling advantage to the Darwinian struggle. Their new type of photosynthesis produced molecular oxygen, a gas toxic to their competitors. This was a poison, a polluting gas never before seen in such quantities on Earth. Microbial gas warfare! Overwhelmed by the new gas, the anaerobic parents could neither fix nitrogen nor manufacture bacteriochlorophyll. Their ability to make the bare necessities of life—proteins, nucleic acids, and even the pigments on which their anoxygenic photosynthesis depended—was suddenly in jeopardy. Only one way out: Retreat or die. Cyanobacteria became kings of the pond.

If this actually happened, wouldn't anoxygenic photosynthesizers have become extinct long ago? No, they did survive. These primitive photoautotrophs had originated and evolved in an anoxic world bathed in deadly UV light. Their photoautotrophy had placed them in a life-threatening situation. To photosynthesize, they had to see the sun; but if they lived where too much light penetrated, they would have been fried alive. The most successful strains therefore developed the ability to glide, to move away from intense UV light or other harmful substances. When toxic-oxygen-spewing cyanobacteria entered the scene, nonmotile anaerobes may well have died in droves, but anaerobic photoautotrophs weathered the storm by retreating to a more agreeable setting. Their ability to move was of **selective advantage**—in fact, it saved them from extinction.

Today, cyanobacteria and anoxygenic photoautotrophs live harmoniously in layered, commonly mound-shaped communities known as **stromatolites**. How can they coexist? The answer lies in their differing light-harvesting pigments. The oxygen-producing cyanobacteria live in the uppermost layers of stromatolites, with the non-oxygen-producing photosynthesizers just beneath. Much of the light energy is absorbed by the cyanobacteria (Figure 4.17), but this does not snuff out the anoxygenic photosynthesis of the green sulfur and purple bacteria that live below because these more primitive photosynthetic anaerobes are literally able to see through the cyanobacterial layer—their pigments absorb light unused by the cyanobacteria above, light that seeps through to power their non-oxygen-producing photosynthesis. Moreover, the gaseous toxic oxygen produced by cyanobacteria in the layer just above these anaerobic microbes rises into the atmosphere, rather than settling where they live; thus, it has little effect on their ability to fix nitrogen or to manufacture bacteriochlorophyll. Primitive (but clever!) anoxygenic photosynthesizers survived by fleeing the war zone. Unable to compete with cyanobacteria, they retreated into a different habitat, one in which they could still absorb life-giving solar energy but in a zone unspoiled by toxic oxygen.

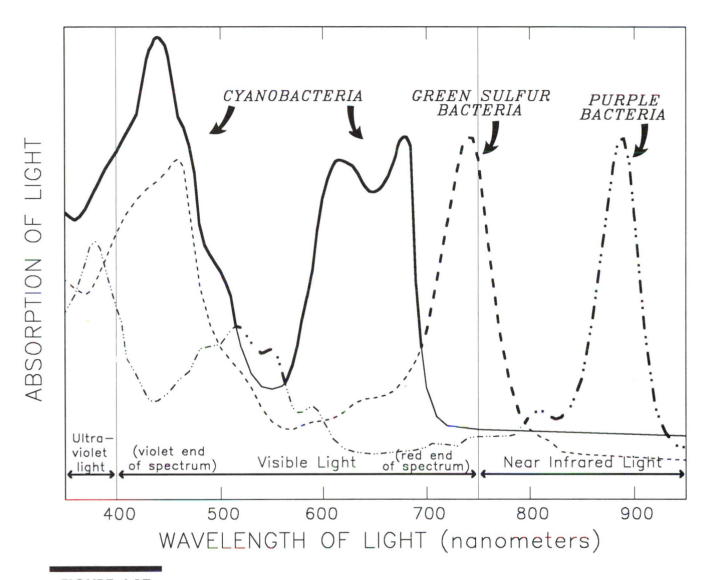

FIGURE 4.17

Oxygen-producing cyanobacteria and non-oxygen-producing green sulfur and purple bacteria absorb light energy from different parts of the spectrum. In stromatolitic microbial communities, anoxygenic photosynthetic bacteria can therefore absorb light energy unused by the oxygen-producing cyanobacteria that inhabit the layer above them.

WHY DO WE BREATHE OXYGEN?

As the old saying goes, "One man's trash is another man's treasure." To anaerobes, oxygen is not only some other organism's waste, it is a poison to be avoided at all costs. To plants, animals, virtually all eukaryotes, just the reverse is true—oxygen is an elixir, the essence of life. Why this disparity? The key is that aerobes not only tolerate oxygen, but require it to fuel their energy-generating process, aerobic respiration. Simply put, they need oxygen to breathe! What is aerobic respiration? How does it work? Where did it come from?

The Nature and Evolutionary Origin of Aerobic Respiration

As shown in Figure 4.18, aerobic respiration is a three-part process. First, *glycolysis* is used to convert glucose to pyruvate, yielding two ATPs (and water) for each molecule of glucose metabolized. Second, in the presence of oxygen, the pyruvate thus produced is broken down by a cyclic process (the *citric acid cycle*) to yield two more ATPs, electrons, and carbon dioxide. Third, as molecular oxygen is added, the electrons generated by the citric acid cycle are conveyed along a series of *electron carriers* to yield 32 additional ATPs. Thus, in the presence of oxygen, breakdown of each molecule of glucose to water and carbon dioxide yields a total of 36 ATPs (summarized in the overall reaction shown at the top of Figure 4.18).

In comparison with earlier-evolved glucose fermentation (that is, glycolysis alone), aerobic respiration provides a vastly improved mechanism of energy production. Whereas

---

FIGURE 4.18

Aerobic respiration is a three-part process: Glycolysis, followed by the citric acid cycle, followed by an energy-yielding (and oxygen-requiring) electron carrier system. If sufficient oxygen is not available, the pyruvate produced by glycolysis is converted to lactic acid and then to glucose, and is recycled into the respiratory pathway.

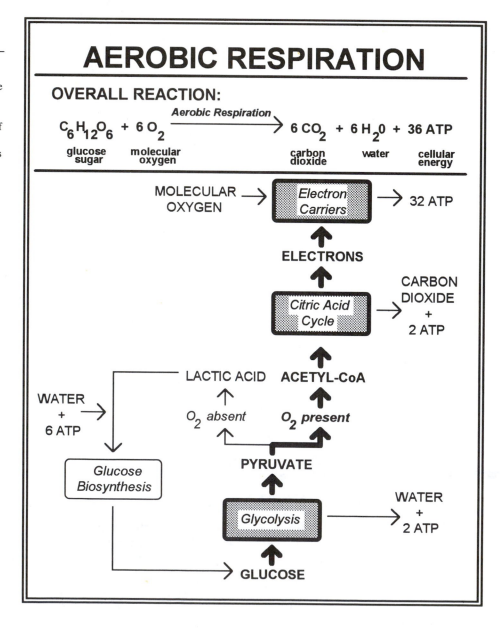

the primitive process yielded only two ATPs per glucose molecule, releasing only 2% of the energy stored in glucose, the new oxygen-using system yielded 36 ATPs, a whopping 38% of the energy potentially available (an energy yield that compares favorably with the 25% efficiency typical of automobile engines). Without doubt, the evolutionary development of aerobic respiration was cost-effective.

How did this tripartite system arise? The first part—glycolysis—was not new with aerobes; inherited from primitive anaerobic heterotrophs, its evolutionary history long predated the advent of aerobic forms of life. Similarly, the second part of the system—the citric acid cycle—also was not new; it was hidden in a different, earlier-evolved form. The citric acid cycle (Figure 4.16, right column) evolved by *reversal* of the cyclic portion of the dark reactions of anoxygenic photosynthesis (Figure 4.16, left column). The third part, the oxygen-consuming electron transport system was also not new. It resembles the electron carrier system that links Photosystems I and II in cyanobacterial oxygenic photosynthesis (Figure 4.15). Once again, *the evolutionary process was conservative and economical,* recasting and reusing earlier-evolved metabolic inventions. However, in addition to depending on previous development of these reusable metabolic mechanisms, the origin of aerobic respiration hinged also on the availability of molecular oxygen. It thus could not, and did not develop until oxygenic photosynthesis had evolved. Clearly, evolution is sequential, a process that operates by remodeling and building on whatever systems already exist. *Past evolution plays a major role in molding the evolutionary future.*

Interestingly, the evolutionary history of aerobic respiration also plays a role in human behavior. Probably like you, when I was a child, I played outside and did a lot of running. After awhile, if I ran fast and hard, especially for a long distance, I began to hurt. I would get what we kids called a "side ache" and I had to stop to catch my breath. That worked; the hurt stopped. Why? The reason is shown in Figure 4.18. My lungs were not bringing in oxygen fast enough to operate the entire three-part sequence of aerobic respiration. Because there was not enough oxygen, the pyruvate produced by glycolysis, the first part of the process, built up faster than it could be used, and it was then automatically diverted to produce ache-causing lactic acid. Slowing down and catching my breath simply gave my body a chance to restore the proper balance between oxygen inflow and pyruvate production; given enough time, the excess lactic acid would be used up by glucose biosynthesis and recycled into respiration (Figure 4.18). Thus, when internal oxygen supplies become exhausted by rapid exercise, we humans (like all other animals) are forced to obtain energy solely from glycolysis and to recycle excess lactic acid by the anaerobic mechanisms of glucose biosynthesis. In short, our bodies resort to the most primitive of all forms of metabolism, ancient processes invented by anaerobic heterotrophic microbes billions of years ago.

SUMMARY: THE FOUR-STAGE DEVELOPMENT OF MODERN METABOLIC PROCESSES

Before considering fossil evidence that can tell us when in the geologic past the various CHON- and energy-yielding metabolic pathways became established, let us briefly review their evolutionary relations (summarized in Figure 4.19). Although some details regarding the relations among these metabolic processes are a bit uncertain (it has been speculated, for example, that anaerobic photoautotrophy might actually be more primitive than anaerobic heterotrophy), most workers agree that they separate into the following four stages of an evolutionary continuum (Figure 4.3).

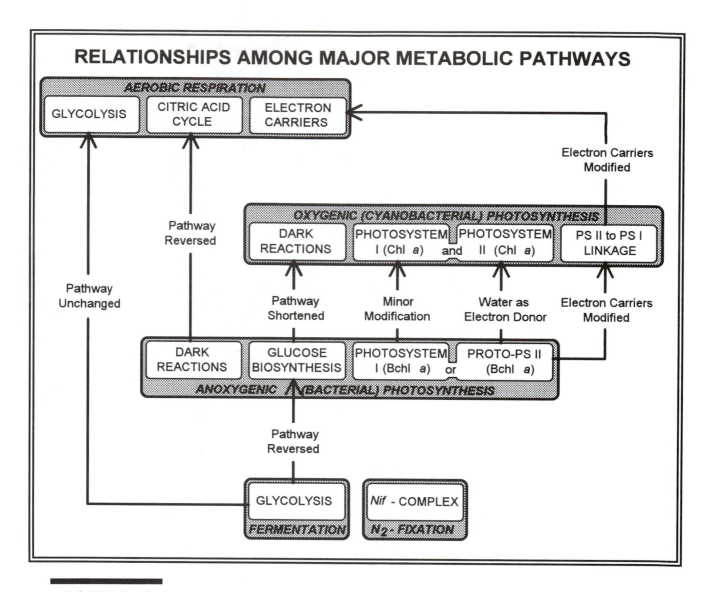

FIGURE 4.19

Evolutionary relationships among major metabolic pathways. As explained in the text, new metabolic inventions have evolved by modification and reuse of earlier-developed metabolic pathways. *The evolution of metabolism has thus been conservative and economical!*

1. Anaerobic Heterotrophy (Glycolysis and N₂-Fixation)

Among the earliest metabolic processes was glycolysis, a simple pathway used by primitive anaerobic heterotrophs to generate energy from breakdown of glucose (Figure 4.3, bottom; and Figure 4.6, left column). As supplies of the glucose fuel (provided initially by abiotic syntheses) became increasingly limited on primordial Earth, a second pathway evolved by reversal of the mechanisms of glycolysis—glucose biosynthesis (Figure 4.6, center column). Glucose biosynthesis provided a new source of CHO, but it was not cost-effective: More energy had to be supplied to manufacture the glucose than was yielded by its biochemical breakdown. A problem was also posed by the scarcity of biologically usable nitrogen in the form of ammonia and nitrate. This difficulty was overcome by invention of the crucial but energy-expensive process of ferredoxin-driven N_2-fixation.

2. Anaerobic Photoautotrophy (Anoxygenic Bacterial Photosynthesis)

Light energy harvested by pigment systems newly evolved in anaerobic heterotrophs provided improved means for photoheterotrophic uptake of CHON-containing compounds from the environment. Modification of this bacteriochlorophyll-based system permitted use of CO_2 as a carbon source and development of anoxygenic (bacterial) photoautotrophy. In this new process (Figure 4.16, left column), electrons generated by light-driven reactions of a single photosystem were fed into a cyclic reaction system to produce the all-important cellular fuel, glucose sugar.

The harnessing of light energy by anaerobic photoautotrophy was a major advance over the preceding "dark ages" in the history of life. For the first time, an energetically cost-effective source for production of ample supplies of glucose had been invented, and the biota was freed from its dependence on nonbiologically made foodstuffs. Nevertheless, these primitive photosynthesizers may not have been widespread; the sources of hydrogen used in this type of photoautotrophy (principally H_2 or H_2S) were only locally abundant, and in the oxygen- and, thus, ozone-deficient early environment, distribution of these earliest photosynthetic microbes would have been restricted by the high influx of noxious UV light penetrating to the surface of the planet.

3. Aerobic Photoautotrophy (Oxygenic Cyanobacterial Photosynthesis)

A more complicated type of photosynthesis evolved next, one based on two interconnected chlorophyll *a*-containing photosystems and which used water as the source of hydrogen. Not only was this hydrogen source plentiful and widespread, but its use resulted in production of gaseous oxygen, a by-product poisonous to the anaerobic photoautotrophs with which these earliest oxygen-producers competed to see the sun. Virtually all the photoautotrophic machinery of these new oxygen-generating mutants (the first true cyanobacteria) was borrowed from the earlier-evolved anoxygenic photosynthesizers—their dark reactions, light reactions, even the electron carriers that linked together their two separate photosystems—all have evolutionary roots in the systems present in anaerobic (bacterial) photoautotrophs (Figure 4.19). Unable to compete with the newcomers, anoxygenic photosynthesizers retreated into a habitat in which they could absorb solar energy but were not threatened by toxic oxygen. The upper portions of the **photic zone** (the sunlit region in which there is sufficient light energy for photosynthesis) were thus open for exploitation by cyanobacteria and, over time, as molecular oxygen (and, thus, UV-absorbing ozone) built up in the environment, these oxygenic photosynthesizers became dominant, microbial rulers of early Earth.

4. Aerobic Heterotrophy (Aerobic Respiration)

For the first time in the history of the planet, major quantities of molecular oxygen were being pumped into the environment. For living systems, this presented untold opportunities—but also terrible dangers. Oxygen is very reactive; that is, it will chemically combine rapidly with many substances. Wildfires are one obvious result. The heat given off by such conflagrations shows us that reaction with oxygen—**oxidation**, the process of burning—must be a tremendous source of energy. The next evolutionary step, development of the three-part process of aerobic respiration, was life's way to tap this rich energy source. None of the parts of the tripartite system was newly invented by aerobes (Figure 4.19), but by tying the parts together, evolution devised a powerful new way of living. The first part, glycolysis, was borrowed unchanged from the successful primitive system developed much earlier by anaerobic heterotrophs. The second part, the electron-generating citric acid cycle, was a revised (reversed) version of a portion of the dark reactions present in anoxygenic photosynthesizers. The third part of the system, in which oxygen is added and much energy is generated, was simply a modified version of the electron-carrier system that had first evolved in photoautotrophs. Whereas the sugar fermentation (glycolysis by itself) of earlier-evolved anaerobic heterotrophs yielded only two ATPs for each glucose metabolized, the new process yielded 36. This is a huge difference in energy production. Indeed, living anaerobic cells have been shown to need as much as 10 times as much glucose as aerobic cells to do the same amount of work, and anaerobes consume many times their cell weight of glucose in only short periods of time, whereas aerobes do not. Energetically, oxygen-using aerobic respiration was thus a real boon to life—clearly, *its evolution was conservative, economical, and energetically cost-effective.*

The Ancient "Modern" Ecosystem

Establishment of the four metabolic processes—heterotrophy and photoautotrophy, both anaerobic and aerobic (Figure 4.3)—marked the first appearance of the complete **"modern" ecosystem**, the ecological structure of the entire present-day world. Ecologically, whether CHON and energy are cycled between animals and plants, as they are today, or solely between heterotrophic and photoautotrophic microbes, as they were in the distant past, makes no difference. Of course, animals and plants are big and microbes are vanishingly small, but the flow of carbon and of energy are identical, the same metabolic processes are involved, and the same rules and relationships apply. Our present biological world is just a scaled-up version of that first developed by an ancient microbial menagerie.

How ancient is this modern ecosystem? The answer can come only from a single source—the preserved rock record, the sole repository of *direct evidence* of the early history of life.

HOW OLD IS THE MODERN ECOSYSTEM?

Evidence from the Oldest Fossils

The oldest *bona fide* fossils yet discovered, nearly three-quarters the age of Earth, are tiny petrified microbes found in a Western Australian rock unit known as the **Apex chert.** Several of the 11 fossil species detected in this ancient deposit are shown in Figure 4.20. The fossil-bearing **chert** (a type of rock composed mostly of very fine-grained quartz) was sedimented in a narrow seaway between towering volcanoes that from time to time blanketed the shallow seafloor with massive flows of lava. The fossiliferous cherts are sandwiched between two such flows, an older lava, about 3.47 billion years in age, and a younger lava flow, about 3.46 billion years old. Thus, the bed containing the fossil microbes was laid down sometime during the intervening 10 million years, more or less 3.465 billion years ago.

In view of their minute size and very great age, the fossils are astonishingly well preserved. Like all living microbes, they lacked "hard-parts"—they had no shells, bones, or

FIGURE 4.20

Ancient microscopic fossils, about 3.5 billion years old, from the Apex chert of northwestern Western Australia. Many of these cellular filaments, the oldest such fossils now known in the geologic record, resemble cyanobacteria living today. The fossils were photographed in slivers of rock (petrographic thin sections) thin enough to see into by use of a high-powered microscope. Because the petrified fossils are three-dimensional and sinuous, photomontages have been used to show the specimens in Parts A, B, D, and F to J. Except as otherwise indicated (in Parts A, B, and D), the size of each of the fossils is shown by the 10 μm scale in Part F. Parts A, B: *Primaevifilum amoenum*. Parts C and F: *Archaeoscillatoriopsis disciformis*. Part G: *Primaevifilum delicatulum*. Parts H and J: *Primaevifilum conicoterminatum*.

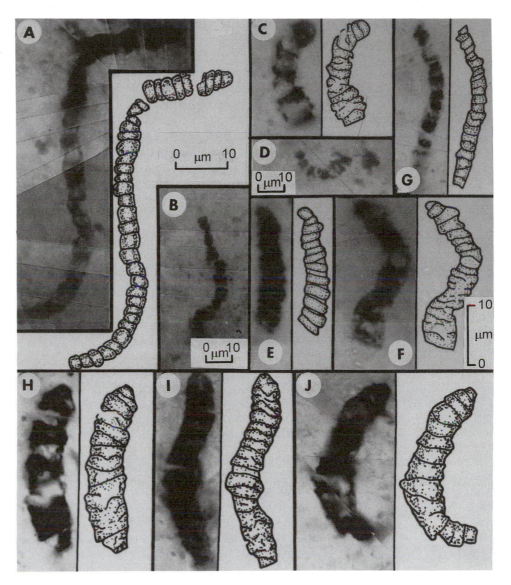

tough, thick-skinned carapaces of the type that lead to easy preservation as many higher forms of life have. Instead, like petrified wood, the fossils are composed only of the coaly remnants of their original carbon-rich cell walls. All their cellular contents (ribosomes, DNA, pigments, and everything else) were leached away long ago. Virtually all that is left to trace their identity is the size, shape, and boxcar-like arrangement of their simple filament-forming cells. However, on the basis of these clues, features that also reveal how the fossil cells divided and multiplied, it seems safe to say that most of the Apex microbes were probably cyanobacteria. Many of the fossil species (and all the specimens shown in Figure 4.20) closely resemble *living* cyanobacteria. In fact, most are similar to members of a particular group of cyanobacteria (known as the Oscillatoriaceae) shown by rRNA evolutionary trees to be among the most ancient of all cyanobacterial lineages. Moreover, the presence of cyanobacteria (or other types of photoautotrophs) in the fossil community is supported by chemical analyses showing that the coaly carbon preserved in the Apex rocks was produced by photosynthesis. Protected from destructive UV light by the shallow water above them, the microbes of this ancient biological community lived within the photic zone where they evidently carried out oxygen-producing photoautotrophy.

Thus, cyanobacterium-like photoautotrophs were present by about 3.5 billion years ago. Because Earth itself is only 4.5 billion years old, these fossils prove that *life began early in Earth history*, within the first billion years of the planet's existence. Moreover, if the Apex fossils include true cyanobacteria, they also prove that during this billion years evolution had advanced very far, very fast. Not only can cyanobacteria cope with reactive oxygen, but all are also capable of both oxygen-producing photosynthesis *and* aerobic respiration. Because both of these are advanced processes that originated by modification of more primitive anaerobic systems (Figure 4.19), their presence would mean that anaerobic heterotrophy and autotrophy must also have existed. Evidently, *all* of the components of the modern ecosystem—anaerobes, aerobes, heterotrophs, and autotrophs—were already in place by this very early stage in the history of life.

It is important to realize that even under the best of circumstances, tiny delicate microbes like those of the Apex chert stand an exceedingly slim chance of being preserved as fossils. Moreover, for these fossils to have survived to the present, they had to have been entombed in rocks that not only escaped mountain building, deep burial, and metamorphic "pressure cooking" (the fate of virtually all ancient geologic units), but that also were not uplifted, eroded, and weathered away during the intervening thousands of millions of years. Yet, survive they did, and although they provide only a scanty glimpse of the total diversity of early life, the fact that 11 different species have been detected in this single fossil horizon is a clear indication that *many* different types of microbes existed at this time in Earth's history. By 3.5 billion years ago, microbial life was diverse, thriving—indeed flourishing—not only in what is now Western Australia but in every other habitable environment over the entire globe.

When Did Life Begin?

The origin of life cannot be dated precisely. Nevertheless, there are well-based notions about when life began. The Apex fossils provide a minimum date; life was present at least 3.5 billion years ago. Moreover, even though these earliest fossil organisms are microbes, they are *advanced* microbes—far too complicated to have been the first forms of life. The origin of living systems must have occurred considerably earlier. Just how much earlier, however, is difficult to know. Because of normal geologic processes (mountain building, weathering, and the like), only one older (severely pressure-cooked) set of rocks, about 3.8 billion years in age, has persisted to the present and no rocks at all appear to have survived from the first 700 million years of Earth's existence. What happened during this earliest, formative phase of Earth history? The Moon, our closest neighbor in space, holds the answer.

Unlike Earth, the Moon lacks an atmosphere. Therefore, it never rains on the Moon and there is no weathering like that on Earth. The pristine lunar surface has retained a record of the period between 4.5 and 3.8 billion years ago, the earliest 700 million years that are missing from the Earth's geology. As documented by the scarred and cratered lunar landscape, this was a horrendous time. The Moon and other bodies of the solar system were in their final stages of formation, sweeping up huge chunks of rocky debris encountered in their orbits—in short, they were being bombarded, blasted, by infalling meteors. Because Earth is so much larger than the Moon, the barrage here was very much greater. In fact, bombardment of Earth was so intense and the infalling meteors so enormous that all the world's oceans were repeatedly vaporized, turned to steam. Any life that had obtained a foothold would have been wiped out. The entire planet would have been sterilized over and over again!

This period of recurring planetary sterilization continued until the last of the huge orbiting chunks had been swept away, about 3.9 billion years ago. Life as we now know it could therefore not have come into being until that time. Yet the Apex fossils tell us that living systems were diverse, flourishing, and widespread by 3.5 billion years ago, only 400 million years later. How could life advance so far so fast?

How Could Evolution Proceed So Far So Fast?

There is a crucial difference in science between knowing and guessing. "Knowing" is based on clearcut evidence; "guessing" is based on what we *think* is true, but cannot know for certain because of lack of evidence. The goal of science is to know, not just to guess. At present, because of the absence of direct fossil evidence, the precise reasons for the speed of evolution between 3.9 and 3.5 billion years ago cannot be known. The best we can do is make an educated guess.

We do, however, have one card up our sleeve. We can use the fossil record of later geologic time as a measure of how much evolution occurs within a 400-million-year-long period, for example, the most recent 400 million years. Near the beginning of this period, the earliest land plants, tiny twigs only a centimeter or so in height, first made their appearance. Over the following 400 million years, plant life evolved step-by-step to produce lush lowland marsh vegetation, the giant spore-producing trees of the Coal Swamp floras, highland seed plants, luxuriant conifer forests and, ultimately, all of the flowering trees, shrubs, and grasses of the modern world. The past 400 million years has also witnessed the evolutionary history of *all* animal life on land—from amphibians (salamanders, frogs, and the like), to reptiles (including dinosaurs), to birds and all the warm-blooded mammals of the present day. Clearly, an enormous amount of evolution can be squeezed into a 400-million-year-long period.

The impressive evolutionary feats of the most recent 400 million years, however, may not be a fair test of early evolutionary rates. By 400 million years ago, life had already been long established, but if life existed 3.9 billion years ago, it was in an embryonic stage. How could *primitive* life have advanced so far so fast? The key is in the mechanisms of early evolution. As we have seen, genes in primitive microbes were duplicated frequently, a surefire mechanism for rapid evolution. Moreover, all of the most important steps in life's early development involved invention of new metabolic capabilities, and almost none of these was totally "new." Rather than starting from scratch, evolution took many shortcuts—metabolism evolved by modification and reuse of successful systems that already existed (Figure 4.19). Thus, combined with the means for rapid evolution provided by gene duplication, *the conservatism and economy of the evolutionary process seem certain to have speeded early evolutionary advance.*

THE SUBSEQUENT HISTORY OF LIFE

Rules of the Evolutionary Game

Strategies used to compete and win in the evolutionary game have changed, markedly, over the long course of life's history. We are familar with the latest chapters of the evolutionary story, those recorded in fossils and sediments of the **Phanerozoic Eon**, the most recent portion of geologic time (approximately 545 million years). When we think of evolution, we think of the history of Phanerozoic life—the progression from seaweeds to flowering plants, from trilobites to humans—a history of relatively rapidly evolving sexually reproducing plants and animals that owe their success to specialized organ systems (flowers, leaves, teeth, limbs) used for exploitation of particular environments. In short, we think of normal evolution played by the "normal rules" of the game.

However, we are much less familiar with earlier events in the Darwinian struggle, those occurring during the **Precambrian Eon**, the nearly 4 billion years spanning the time from the origin of the planet to the beginning of the Phanerozoic. This earlier and very much longer phase of evolution featured a different cast of characters playing the game by a more primitive set of rules. In place of multicellular plants and animals, the Precambrian biota was composed largely of simple nonsexual prokaryotes. Rather than evolving rapidly, these Precambrian microbes evolved at an extraordinarily slow (**hypobradytelic**) rate. Instead of having specialized organ systems for exploitation of particular environments, the most successful of these early-evolving microorganisms were ecologic generalists, able to withstand the rigors of an exceptionally broad range of harsh environments.

In addition to these differences in evolutionary strategy, there is at least one other striking contrast between Precambrian and Phanerozoic life—unlike their primitive predecessors, some Phanerozoic organisms were intelligent. How did this come about?

The Origin of Intelligence

It is assumed in some quarters that intelligence is an exclusively human trait. Almost certainly, however, this is not correct. Chimpanzees, dolphins, dogs, cats, horses—they all seem to think. Even pigs are claimed to be surprisingly smart. What about an eagle, for example? How does it catch its prey? It doesn't swoop to where the rabbit *was*, it dives to where the rabbit *will be* when the eagle gets there to snatch it. Doesn't that require forethought? What about a frog, patiently lying in wait for its insect meal, or a fiddler crab, scurrying about and holding up its claw to attract mates and discourage other crabs from entering its home territory? These organisms certainly seem to know what *is*—and even to foresee what *will be*—going on about them. How much of this reflects intelligence?

Definitions of intelligence vary. However it is defined, most agree that intelligence is shown by the way an organism interacts with and alters its environment, and that the roots of intelligence lie somewhere within the evolutionary Tree of Life. Why is it that some organisms are smart and others evidently are not?

Clearly, plants do not qualify as intelligent. Similarly, it is hard to believe that nonmotile animals like sponges, corals, sea lilies, and the like are particularly smart—all they seem to do is sit in one place and "vegetate." Perhaps that's the answer. If an organism does not need to move from one place to another—for instance, a tree or shrub, or the builders of a coral reef—it is unlikely to need much intelligence to live. Its basic requirements for CHON and energy can be met fully by a sedate mode of life (that is, by passively collecting solar energy, like a tree, or by waiting for food particles to settle into its environs, like a coral).

However, this is not true of *mobile* multicellular heterotrophs (Figure 4.21). Mobility requires energy, and these organisms need much more nourishment—larger quantities of CHON and energy—than sessile forms of life. Efficient foraging, however, demands a carefully crafted strategy. It seems plausible, therefore, that the beginnings of what we regard as intelligence are tied to the early evolution of feeding in mobile eukaryotic heterotrophs.

Among early-evolving mobile multicellular heterotrophs (discussed in Chapter 3), those best able to ferret out foodstuffs were most successful. Food-seeking and food-engulfing apparatus thus became localized at the "encounter" end of these heterotrophs, resulting in a front-to-rear organization that has been carried over to all of their descendants (including humans). To be efficient, nerve systems needed to drive the food-ferreting apparatus

FIGURE 4.21

Major steps in the evolution of
intelligence in mobile multicellular
heterotrophs.

WHY ARE HIGHER ANIMALS SMART?

INTELLIGENCE

CEPHALIZATION

SENSORY APPARATUS AT "ENCOUNTER END"

MOBILITY WITH FRONT/REAR ORGANIZATION

INTENSIVE MOBILE FEEDING

HIGH "CHON" AND ENERGY REQUIREMENTS

MOBILE MULTICELLULAR HETEROTROPHY

also became localized at the encounter end, but concentration of these critical systems in a single area increased their susceptibility to life-threatening damage. For this reason, mobile heterotrophs that had a protective thick skin or carapace overlying this vulnerable region were most successful, and their success led to increased **cephalization**, development of a "hardified" head region protecting the neural equipment necessary to process complex sensory information. Over time, the sensory apparatus and associated neural equipment became progressively more complicated and refined, resulting in evolutionary development of increasingly intelligent forms of life.

Viewed this way, intelligence is a logical result of the evolution of mobile heterotrophy. Mobile heterotrophy, however, developed simply as an effective way to satisfy the necessities of life—the need for CHON and energy. And these necessities as well as the two strategies that evolved to meet them, autotrophy and heterotrophy, are products of much earlier evolution, set in motion billions of years ago when the planet was very young. Thus, the seeds that ultimately flowered in human intelligence can be found in the ecologic structure of Earth's earliest biosphere. Housed in our cells, in the chemistry that keeps us alive, we each contain metabolic memories of this unimaginably long evolutionary journey. From microbes to man, from ignorance to intelligence, _the evolutionary process has remained remarkably conservative and economical._

■

FURTHER READING

Evolution of Metabolic Pathways

Broda, E. 1975. *The Evolution of Bioenergetic Processes* (New York: Pergamon Press).

Schopf, J.W. 1992. Times of origin and earliest evidence of major biologic groups. In: J.W. Schopf and C. Klein (Eds.), *The Proterozoic Biosphere, A Multidisciplinary Study* (New York: Cambridge Univ. Press), pp. 587–593.

rRNA Phylogeny

Iwabe, N., Kuma, K., Hasegawa, M., Osawa, S., and Miyata, T. 1989. Evolutionary relationships of archaebacteria, eubacteria, and eukaryotes inferred from phylogenetic trees of duplicated genes. *Proc. Nat. Acad. Sci. USA 86:* 9355–9359.

Rivera, M.C., and Lake, J.A. 1992. Evidence that eukaryotes and eocyte prokaryotes are immediate relatives. *Science 257:* 74–76.

Woese, C.R. 1987. Bacterial evolution. *Microbiol. Rev. 51:* 221–271.

Woese, C.R., Kandler, O., and Wheelis, M.L. 1990. Towards a natural system of organisms: Proposal for the domains Archaea, Bacteria, and Eucarya. *Proc. Nat. Acad. Sci. USA 87:* 4576–4579.

Gene Duplication

McLachlan, A.D. 1987. Gene duplication and the origin of repetitive protein structures. *Cold Spring Harbor Symposia on Quantitative Biology LII:* 411–420.

Sonti, R.V., and Roth, J.R. 1989. Role of gene duplications in the adaptation of *Salmonella typhimurium* to growth on limiting carbon sources. *Genetics 123:* 19–28.

Photosynthesis

Blankenship, R.E. 1992. Origin and early evolution of photosynthesis. *Photosyn. Res. 33:* 91–111.

Olson, J.M., and Pierson, B.K. 1987. Evolution of reaction centers in photosynthetic prokaryotes. *Internat. Rev. Cytol. 108:* 209–248.

Evolution of Ferredoxin

Eck, R.V., and Dayhoff, M.O. 1966. Evolution of the structure of ferredoxin based on living relics of primitive amino acid sequences. *Science 152:* 363–366.

The Primitive Environment

Abelson, P.H. 1966. Chemical events on the primitive Earth. *Proc. Nat. Acad. Sci. USA 55:* 1365–1372.

Schopf, J.W. 1978. The Precambrian development of an oxygenic atmosphere. In: *Genesis of Uranium- and Gold-Bearing Precambrian Quartz-Pebble Conglomerates, U.S. Geological Survey Professional Paper 1161-B:* B1–B11.

Schopf, J.W. (Ed.). 1983. *Earth's Earliest Biosphere, Its Origin and Evolution* (Princeton: Princeton Univ. Press).

Earliest Known Fossils

Schopf, J.W. 1992. Paleobiology of the Archean. In: J. W. Schopf and C. Klein (Eds.), *The Proterozoic Biosphere, A Multidisciplinary Study* (New York: Cambridge Univ. Press), pp. 25–42.

Schopf, J.W. 1993. Microfossils of the Early Archean Apex chert: New evidence of the antiquity of life. *Science 260:* 640–646.

Precambrian and Phanerozoic Evolution

Schopf, J.W. (Ed.) 1992. *Major Events in the History of Life* (Boston: Jones and Bartlett).

Schopf, J.W. 1992. A synoptic comparison of Phanerozoic and Proterozoic evolution. In: J.W. Schopf and C. Klein (Eds.), *The Proterozoic Biosphere, A Multidisciplinary Study* (New York: Cambridge Univ. Press), pp. 599–600.

Schopf, J.W. 1994. Disparate rates, differing fates: Tempo and mode of evolution changed from the Precambrian to the Phanerozoic. *Proc. Nat. Acad. Sci. USA 91:* 6735–6742.

CHAPTER 5

HOMEOTIC GENES AND THE EVOLUTION OF BODY PLANS

■

E. M. De Robertis*

EMBRYONIC DEVELOPMENT AND EVOLUTION

It has recently been discovered that the development of body form in animals is controlled by a gene network that is seemingly universal, common to all types of animals. This chapter reviews the impact that this discovery is having on our current view of the evolutionary process.

In 1894, William Bateson (1861–1926), a famous British naturalist, attempted to explain Darwinian evolution using as a clue **homeotic transformations** (malformations in developing or regenerating animals in which a segment or region of the body is transformed into the likeness of some other normal segment). In 1923, the American geneticists Calvin Bridges (1889–1938) and Thomas H. Morgan (1866–1945) discovered the *bithorax* mutation in fruit flies. The extensive genetic analysis of the part of the chromosome in which this mutation occurs was developed by another American geneticist, Edward Lewis, in 1947, and this research is still continuing (Lewis, 1978). Mutations in genes of the *bithorax* complex, together with those of a second group of genes, the *Antennapedia* complex, lead to homeotic transformations in the common fruit fly, *Drosophila*. Study of the genes causing these homeotic malformations in flies showed that they have a region in which the DNA sequence is very similar (that is, it has been "highly conserved" during evolution) (Gehring, 1994). Discovery of this relatively constant stretch of sequence, the **homeobox sequence**, was the key that permitted discovery of related genes in other organisms (termed *Hox* genes), first in frogs (Carrasco et al., 1984) and eventually in many types of multicelled animals.

Homeobox genes are important because their discovery opened the door for understanding the way embryos develop in higher organisms, such as vertebrates, in which it is not possible to carry out the type of genetic analyses that are done routinely in the laboratory on fruit flies. The development of chemical probes with which to search out and identify *Hox* genes in higher animals has caused a revolution in our understanding of em-

*Molecular Biology Institute and Department of Biological Chemistry, University of California, Los Angeles, CA 90095.

bryonic development. Together, genetics and embryology now help us understand how the enormous variety of body forms arose during the hundreds of millions of years of animal evolution.

WILLIAM BATESON AND THE STUDY OF HOMEOTIC TRANSFORMATIONS

Studies of evolution and embryology have been intertwined to varying degrees throughout the history of modern science (Raff and Kaufman, 1983). Naturalists such as William Bateson were intrigued by the question of how new body structures were generated in the course of evolution, and to this end they analyzed malformations occurring spontaneously in animals in the wild. Studies of such "freaks of nature" have added greatly to human knowledge. One example of the malformations that Bateson studied is shown in Figure 5.1, in which the leg of a moth has become transformed into an extra wing. Bateson was interested in changes of one region of the body into another—changes now known as homeotic transformations—because he wanted to explain why new species often seem to arise by abrupt, evidently discontinuous changes in body form, rather than by a series of small, nearly imperceptible alterations. In 1894, Bateson wrote:

> Upon the accepted view it is held that the Discontinuity of Species has been brought about by a Natural Selection . . . [acting on] a continuous series of variations. . . . The difficulties besetting this doctrine . . . have oppressed all who have thought upon these matters . . . and they have caused anxiety even to the faithful.

But Bateson thought there was a way around this problem if mutations could be shown sometimes to affect large regions of the body all at once, causing immediate, extensive changes in body form. Thus, in Bateson's view:

> The case of the modification of the antenna of an insect into a foot, of the eye of a crustacean into an antenna, of a petal into a stamen, and the like, are examples of the same kind. It is desirable and indeed necessary that such variations, which consist in the assumption by one member of a meristic series [a series of body segments] of the form or characters proper to other members of the series, should be recognized as constituting a distinct group of phenomena. . . .

FIGURE 5.1

Drawing of a moth (*Zygoena filipena*), having an extra wing (indicated by the arrow) that has replaced a leg. Transformations like this, of one region of the body into the likeness of another, attracted the attention of William Bateson who sought to prove that the discontinuity between species is the result of the discontinuity of natural variation. (Modified from Bateson, 1894.)

> I therefore propose . . . the term Homeosis . . . for the essential phenomenon is not that there merely has been a change, but that something has been changed into the likeness of something else.

Bateson further asserted that:

> The discontinuity of species results from the discontinuity of variation. Discontinuity results from the fact that bodies of living things are made of repeated parts. Meristic variation in numbers of parts is often integral, and thus discontinuous . . . [A structure] may suddenly appear in the likeness of some other member of the series, assuming at one step the condition to which the member copied attained presumably by a long course of Evolution.

Homeosis in Humans

Bateson—today acclaimed for his pioneering studies of heredity and as the person who coined the term "genetics" (which we now know is based on phenomena that occur in chromosomes at the submicroscopic, molecular level)—was in his day renowned for his studies of the skeletons of large lumbering sloths. As shown in Table 5.1, there is considerable variation in the number of backbone parts termed **thoracic vertebrae** among individual three-toed sloths (*Bradypus*). Thoracic vertebrae are those that occur in the thoracic region of the body and that therefore each have a pair of ribs attached; in three-toed sloths, the number of these vertebrae varies from 14 to 16. In contrast, Bateson found that two-toed sloths (*Choloepus*), members of a different but closely related genus, have 23 or 24 thoracic vertebrae. Evolution from one genus to the other involved a huge change in the number of ribs, showing how a body plan can be altered in discontinuous jumps.

TABLE 5.1

Number of vertebrae in individual animals of three-toed (*Bradypus*) and two-toed sloths (*Choloepus*).*

	Type of Vertebra			
Type of Sloth	Cervical (Neck)	Thoracic (Upper Back)	Lumbar (Lower Back)	Sacral (Pelvic)
Three-Toed	8	15	3	7
Three-Toed	9	14	4	5
Three-Toed	9	15	4	5
Three-Toed	9	15	4	5
Three-Toed	9	15	4	5
Three-Toed	9	15	4	6
Three-Toed	9	15	4	6
Three-Toed	9	15	4	6
Three-Toed	9	16	3	6
Two-Toed	7	23	3	7
Two-Toed	7	23	3	8
Two-Toed	7	23	4	5
Two-Toed	7	24	3	7

*Determined by William Bateson (1894).

FIGURE 5.2

Homeotic transformations occur in humans. This skeleton has extra ribs (shown by shading) in the neck region, attached to the seventh cervical vertebra. Such cervical ribs occur commonly in human populations and are usually not deleterious. (Modified from Kühne, 1932.)

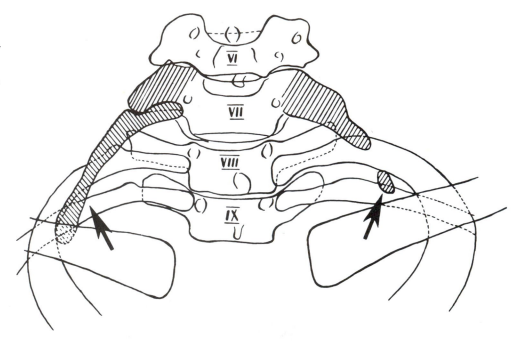

There are also variations among humans. For example, extra ribs sometimes occur in the neck region (as shown in Figure 5.2, attached to the seventh cervical vertebra). Although most people have 12 pairs of ribs, some have one less and some have one more, especially in the lower back (lumbar) region (Table 5.2); when a vertebra in this region acquires a pair of ribs, as shown in Figure 5.3, the pair by definition is considered to be thoracic. In humans, extra ribs occur among members of particular families; the trait appears to be inherited, passed genetically from parents to their children (Figure 5.3). Thus, an extra rib in humans, and in other mammals as well, constitutes a true homeotic transformation in exactly the sense proposed long ago by Bateson.

DROSOPHILA HOMEOTIC MUTANTS AND THE HOMEOBOX

The genetic basis of homeotic transformations was first established in the common fruit fly, *Drosophila* (Bridges and Morgan, 1923; Lewis, 1978). One of the most famous of

TABLE 5.2

Variations in the number of ribs detected by x-rays in a normal human population.

Number of Pairs of Ribs	Rib-Bearing Vertebrae	Male (%)	Female (%)
11	VIII to XVIII	0.8	3.85
12	VIII to XIX	92.1	93.00
13	VIII to XX	7.1	3.10

Source: Data from Kühne, 1932.

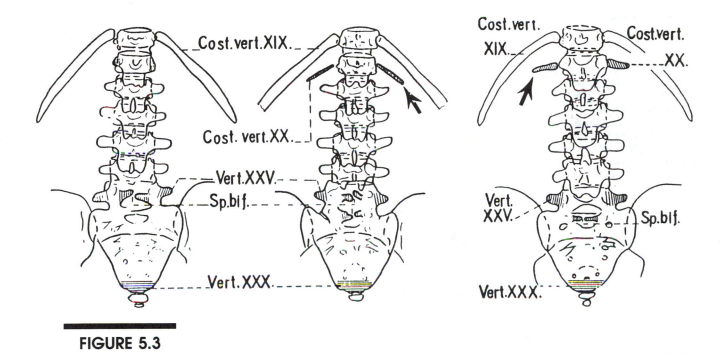

FIGURE 5.3

Human extra ribs have a genetic basis. The vertebral columns of three brothers are shown, two of which have an extra (13th) rib (indicated by arrows) rather than the usual 12. Their father also had 13 ribs, and their mother had an exceptionally long 12th rib (not shown). Because extra ribs like this occur rarely in the general population, their occurrence in the same family indicates that this is probably a heritable characteristic. (Modified from Kühne, 1936.)

these mutations is called *bithorax,* in which an organ that serves as a balancer during flight (the haltere organ) is mutated into a winglike structure (Figure 5.4). Because the making of a wing is complicated, no doubt requiring hundreds of gene products, it was recognized early that the *bithorax* mutation must involve a change in a **selector gene** that controls the activities of many other subordinate genes. The genetic studies of Lewis (1978) showed that the *bithorax* is in fact part of a complex of homeotic genes that occur in a linear order in the DNA of the chromosomes in exactly the same order as the individual body segments that they affect. This interesting and unexpected property—the **colinearity** of homeotic genes and the products of their activity, the body segments that they affect—is discussed in detail later. Scientists have long been interested in homeotic genes, and when it became possible in the early 1980s to **clone genes** (to make many copies of a given gene in the laboratory, so that its chemistry and function can be studied more easily), several groups of scientists were ready to isolate and study the homeotic genes.

A crucial discovery occurred when one of the groups of homeotic genes thus cloned in *Drosophila,* the *Antennapedia* complex, was found to contain a sequence of DNA that is very similar to that occurring in other genes of these fruit flies (McGinnis et al., 1984; Scott and Weiner, 1984). Because of this similarity, these DNA sequences from two genes can bind to each other—that is, they can **hybridize** by linking together through chemical bonds (an exceptionally useful property, discussed also in Chapters 2 and 6). Therefore, using radioactively labeled *Antennapedia* DNA to hybridize with, it is possible to find and essentially "fish out" other genes from the chromosomes that contain this same region is possible. The portion of the *Antennapedia* complex that binds to other genes, and the corresponding portion of those genes that bind to the *Antennapedia* complex, has been

FIGURE 5.4

The *bithorax* homeotic mutation in the fruit fly, *Drosophila*, as originally depicted by Bridges and Morgan. (a) A mutant fly with an extra pair of wings. (b) A normal fly showing the balancer (haltere) organ (at arrow). (c to g) Various degrees of transformation of the balancer organ into winglike structures. (Modified from Bridges and Morgan, 1923.)

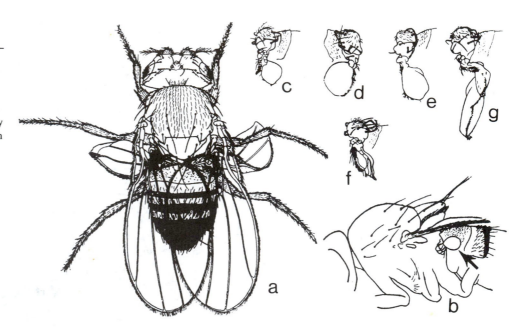

designated the homeobox (McGinnis et al., 1984). Many of the new genes isolated in this way corresponded to chromosomal sites of previously identified homeotic mutations (such as those named *Ultrabithorax* and *Deformed*).

Such DNA sequences bind to each other because they are similar, even though they are in different genes, and they are similar because the genes share a common gene ancestor—they have changed relatively little since they originated by modification of an original gene containing a prototype homeobox sequence. The homeobox for this reason is said to be **conserved in evolution**, little changed since it originated hundreds of millions of years ago. The reason for this very convenient sequence conservation is that homeobox-containing genes carry the chemical code that is used to manufacture proteins that bind to DNA. As shown in Figure 5.5, the conserved region corresponds to the DNA-binding portion of the protein known as the **homeodomain** (the part of the chromosomal gene that contains the code for the manufacture of this protein is called the **homeobox**). The conserved part of the protein, the homeodomain, is little changed during evolution because for it to bind to DNA it must fit, like a key in a lock, into the major deep groove of helically wound molecules of DNA. If this critical part of the protein were to change through evolution, the key would not fit the lock and the protein would not do its job. Other regions of these DNA-binding proteins are not nearly so conserved as is the homeodomain portion produced by the homeobox. That this part of homeotic genes has changed little over the long course of animal evolution is truly fortunate.

HOMEOBOXES IN VERTEBRATE ANIMALS

With the benefit of hindsight, perhaps it should not have been surprising that all higher animals, the multicelled **metazoans**, share similar mechanisms that specify where particular groups of early-forming cells end up as the organism develops and grows into an adult. Without exception, all **eukaryotic** (nucleus-containing) **cells**, the building blocks of every advanced form of life on Earth, share a plethora of other common components (structural proteins such as chromosome-associated histones, a great number of biochemical

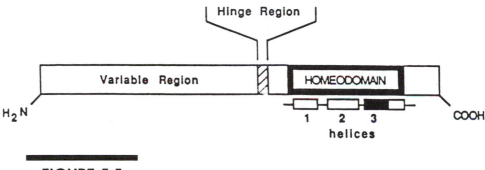

FIGURE 5.5

Homeobox genes code for proteins that bind DNA. The proteins are composed of a variable region, which can differ greatly in its amino acid sequence from one species to another, and a region called the homeodomain (coded for by the homeobox in chromosomes) that chemically binds to specific sequences of DNA and is very similar (evolutionarily conserved) in all multicelled animals. The homeodomains of these proteins consist of 60 amino acids and are composed of three segments (helices) of which helix 3 binds to the major groove of the DNA of chromosomes. Because most homeodomains are very similar in this segment, all bind to similar sites in DNA.

pathways, and various types of intracellular membranes, organelles, and so forth). Nevertheless, when the simple experiment of using the homeobox from the fruit fly to isolate a closely comparable gene from the frog (*Xenopus*) was first carried out (Carrasco et al., 1984), no one would have predicted the remarkable degree of similarity in the molecular mechanisms that control fundamental aspects of the development of body form in these seemingly very distantly related animals.

Soon the monumental task of identifying and isolating all of the homeobox genes that occur in mammals was underway in laboratories throughout the world. Because studies of *Drosophila* showed that homeotic genes are clustered together in the chromosomal DNA, from the start those working on the mouse (*Mus*) were dedicated to showing that homeobox genes were organized in similar clusters in mouse chromosomes, and to mapping their locations on these chromosomes (Rabin et al., 1985). This strategy turned out to be extremely informative. After an extraordinary amount of work, the human and mouse *Antennapedia-* (*Antp-*) type complexes were shown to map into four separate gene complexes, each containing about ten homeobox genes, as is shown in Figure 5.6. The complexes were numbered 1 through 4, and the name **Hox** was coined for genes in the mammalian *Antp*-type complexes. Within the *Hox-1* through *Hox-4* complexes thus named, each gene was initially designated according to the order in which it was discovered.

By 1992, the genes within each vertebrate *Hox* complex could be aligned into 13 separate groups, all of which share distinct similarity in the homeodomains of the proteins they produce (Scott, 1992), and the four originally numbered complexes were renamed in alphabetical order *Hoxa, Hoxb, Hoxc, Hoxd*. Within each of these complexes, the group farthest toward one end of the chromosome (technically, that toward the 3′ end of the DNA molecule, a group related to the *labial* gene in fruit flies) is designated number 1, and the group closest to the other end of the chromosome (the most 5′ group, related to the *Abdominal-B* gene of *Drosophila*) is designated number 13. Thus, to refer to these homeotic genes there is now a nomenclature of *Hoxa-1* through *Hoxa-13*, *Hoxb-1* through

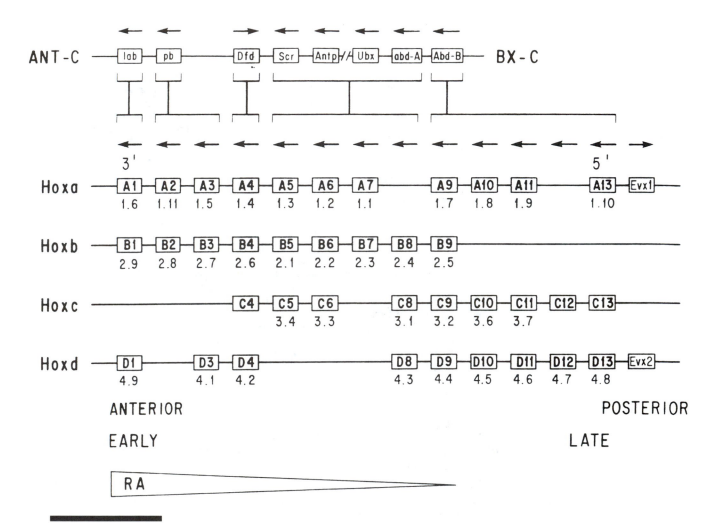

FIGURE 5.6

Vertebrates have four *Hox* complexes, each composed of approximately 10 genes (denoted by boxes) which on the basis of their similarities can be aligned into the 13 groups shown. All *Hox* complexes display colinearity: Genes at one end (the 3′ end) are switched on first, whereas those at the other (5′) end are turned on last, and the proteins produced by the genes at one end of the complex affect the anterior portion of the developing embryo, whereas those at the other end affect the posterior portion. Currently used nomenclature (Scott, 1992) is noted in the boxes (*Hoxa* A1, A2, A3, etc.), and the corresponding earlier-used numerical nomenclature for the mouse (*Hox* 1.6, 1.11, 1.5, etc.) is shown beneath each box. *Drosophila* genes of the *Antennapedia* (ANT-C) and *bithorax* (BX-C) complexes are shown at the top; arrows indicate the direction in which the coded information in the gene is transcribed to form the proteins they produce.

Hoxb-13, and so forth, a system that applies to every type of vertebrate animal, humans included, all of which share the common organization of the *Antp*-type *Hox* complex. There are many other homeobox genes in chromosomes, but most of these have changed considerably since their evolution from the original *Antp*-type ancestral sequence. This chapter concerns only genes of the *Hox* complexes, genes that are especially important because they mediate homeotic mutations.

HOX COMPLEXES

Colinearity in *Hox* Complexes

When the DNA subunits of the mammalian *Antp*-type homeobox were compared with those of their *Drosophila* counterparts (Boncinelli et al., 1988), a remarkable fact became apparent: Homeoboxes at one end of the mammalian *Hox* complexes are similar in their subunit sequence to those at one end of the *Drosophila* homeotic complexes (corresponding to the gene called *labial*), and those toward the other end of the mammalian homeobox resemble the fruit fly *Abdominal-B* gene that also occurs at the other end of its gene complex. The same orientation present in the complexes of fruit flies is present in you, me, your dog, and your cat!

Recently, a cluster of *Antp*-type homeobox genes was identified in a primitive worm, the nematode *C. elegans* (Bürglin et al., 1991; Kenyon and Wang, 1991). As shown in Figure 5.7, nematodes contain homeoboxes related to several fruit fly genes, namely *labial*, *Deformed*, *Antp*, and *Abdominal-B*. Even hydra, a primitive jellyfish-like metazoan, has *labial*, *Deformed*, *Antp*-type, and *proboscipedia* homeoboxes (Murtha et al., 1991; Schierwater et al., 1991; Schummer et al., 1992).

```
Consensus  RKRGRTTYTRYQTLELEKEFHFNRYLTRRRRIEIAHALCLTERQIKIWFQNRRMKWKKEN

Labial     NNS---NF-NK-LT---------------A------NT-Q-N-T-V---------Q--RV
Hoxb 1     PGGL--NF-TR-LT---------K--S-A--V---AT-G-N-T-V---------Q--RE
Ceh-13     NGTN--NF-TH-LT-------TAK-YN-T--T---SN-K-Q-A-V---------E--RE

Deformed   P--Q--A---H-I---------Y---------------T-V-S------------------D-
Hoxb 4     P--S--A---Q-V---------Y---------V---------S-----------------DH
Ceh-15     E--Q-TA---N-V---------THK----K----V--S-M-----V----------H----

Antp       -----Q------------------------------------------------------
Hoxb 7     -----Q-----------------Y---------------T---------------------
mab-5      S--T-Q--S-S----------YHK----K--Q--SET-H-----V---------H---A

Abd-B      VRKK-KP-SKF---------L--A-VSKQK-W-L-RN-Q-----V----------N--NS
Hoxb 9     SRKK-CP--K----------L--M----D-H-V-RL-N-S---V---------M--L-
Ceh-11     S-K--Q--Q----SV--AK-QQSS-VSKKQ-E-LRLQTQ--D------------A---K
```

FIGURE 5.7

Homeobox genes of fruit flies, vertebrates, and nematode worms are very similar and must have originated in a primitive metazoan ancestral to all. *Drosophila* genes listed are *labial, Deformed*, and *Abd-B*; mouse genes are *Hoxb 1, 4, 7,* and *9*; and nematode (*C. elegans*) genes are *Ceh-13, Ceh-15, mab-5,* and *Ceh-11*. Letters indicate particular amino acids occurring in homeodomains coded for by homeobox genes. The list of amino acids shown at the top is the "consensus" sequence, the amino acids that occur most commonly at each position in the *Antp*-type homeodomain. In the four groups of sequences shown below, hyphens indicate the occurrence of the consensus amino acids. Although some amino acids have changed as members of each of these four groups have evolved, the changes within each group are quite similar.

The presence of essentially the same gene complexes in mammals, fruit flies, nematodes, and jellyfish-like hydra means that homeotic gene complexes were already present in some common ancestor long, long ago—that is, in a very primitive, very early-appearing lineage of multicelled animals (such as those discussed in Chapter 3) from which all later metazoans were derived.

The resemblance of *Hox* genes in vertebrates to homeotic genes in fruit flies is not only structural, but also functional. Gene expression studies show that the location of the *Hox* genes in vertebrate animals is colinear with their expression in various body regions (Gaunt et al., 1988; Graham et al., 1989; Douboule and Dollé, 1989). In other words, as is the case in *Drosophila*, genes farthest toward the 3′ end of the complex are expressed toward the front end of the mammalian body, and those at 5′ end are expressed toward the rear, as illustrated in Figure 5.8. This rule of colinearity—of the first to last genes in the complex being expressed front to rear in the organism—applies to all four of the mammalian *Hox* complexes.

The discovery of this colinearity meant that the *Drosophila*-based model for the operation of the system first proposed by Edward Lewis in 1978 also applies to mammals. Thus, the molecular mechanisms that determine the front to rear (**anteroposterior**) axis

FIGURE 5.8

Homeobox genes are clustered in the chromosomes and are expressed in the body in the same order in which they occur in the chromosomal DNA. For example, genes of the *labial* class (*Lab*, upper row, far right) are expressed toward the front end of the fruit fly, whereas genes of the *Abdominal-B* class (*Abd*-B, upper row, far left) are expressed toward the rear. As shown below, the same relations occur in the mouse. The exceptional conservation of these relations suggests that the gene system that controls anteroposterior cell distributions probably arose very early in the evolutionary history of the Metazoa. (Reproduced, with permission, from E. De Robertis et al., *Scientific American*, July 1990, pp. 46–52).

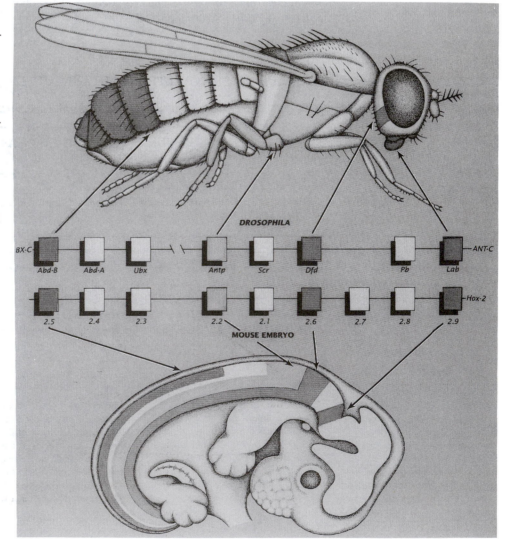

FIGURE 5.9

Mouse tissue section stained to
show the distribution of protein
produced by the *Hoxc 6* gene,
detectable in eight segments (T2
through T9) of the thoracic region.

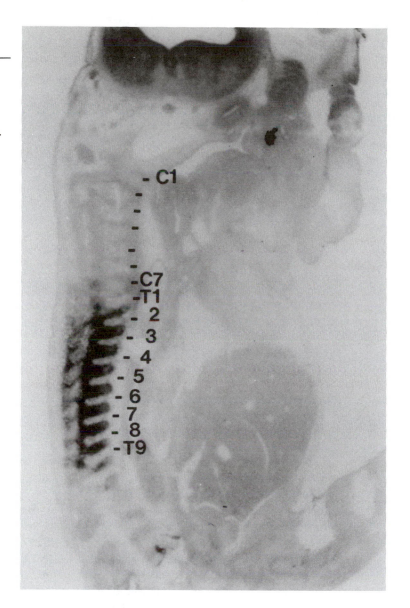

have been conserved in evolution to a degree beyond anyone's expectation (De Robertis
et al., 1990; McGinnis and Krumlauf, 1992). With the discovery of this colinearity in
mammals, the study of homeobox genes suddenly became a hot topic.

The Function of Vertebrate *Hox* Genes

Early in the history of these studies, *Hox* genes were shown to be involved in controlling
the formation of specific regions of the body (Awgulewitsch et al., 1986). Figure 5.9 shows
a mouse embryo stained with a type of biochemical tag called an **antibody** that reacts
chemically with the protein produced by the gene at position number 6 in the *Hoxc* complex (therefore termed the "*Hoxc 6* protein"). *Hoxc 6* is a typical *Hox* protein (of historical interest because it was the first *Hox* gene product to be isolated) that is expressed in
a beltlike band encompassing eight body segments (**somites**) in the thoracic region of the
developing mouse embryo. Thus, *Hox* genes affect the development of regions of the embryo, not just particular tissues or cell types. A main conclusion derived from this regional

FIGURE 5.10

Mouse embryo having extra ribs formed by artificial stimulation of the *Hoxc 6* gene. Normally, the protein produced by *Hoxc 6* affects development only in the thoracic region (T2 through T9, as shown in Figure 5.9). In this mouse, many copies of altered human DNA have been introduced, resulting in production of nine abnormal ribs (indicated by arrows) on vertebrae of the lumbar region. Experiments of this type indicate that vertebrate *Hox* genes can indeed produce homeotic transformations (Jegalian and De Robertis, 1992).

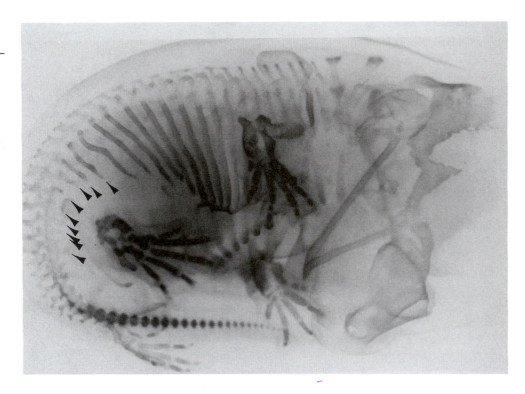

localization is that the vertebrate embryo is partitioned into specific areas, each affected by *Hox* gene activity even before particular tissues or organs are formed.

Notably, the *Hox* proteins from mammals have been found to function also in fruit flies in which they can alleviate the effects of homeotic mutations (Malicki et al., 1990; McGinnis et al., 1990). Inactivation of the *Hoxc 8* gene in the mouse, normally expressed in the thoracic region of the embryo, produces a homeotic transformation that adds ribs to the forward-most lumbar vertebra and converts it into an extra thoracic vertebra (Le Moulleic et al., 1992)—exactly the sort of homeotic transformation that Bateson had foreseen.

The fruit fly studies seemed to predict that the loss of function of a *Hox* gene would result in rear-to-front (posteroanterior) transformations, like that caused by inactivation of the *Hoxc 8* gene. However, this rule does not always apply. When homeodomain-containing proteins are misexpressed in the posterior of a mouse embryo, extra ribs can also be formed in lumbar vertebrae (Jegalian and De Robertis, 1992; Pollock et al., 1992). The mouse embryo shown in Figure 5.10 contains altered human DNA that causes an abnormal *Hoxc 6* protein to occur at high levels in posterior regions in which this protein is normally absent (see Figure 5.9), altering the lumbar vertebrae in such a way that they have acquired nine extra pairs of ribs. Thus, in this example, an overabundance of the protein product of a thoracic *Hox* gene led to formation of thoracic skeletal elements in the lumbar vertebrae, a front-to-rear (anteroposterior) transformation. Each individual cell in the trunk or torso region of the body contains a particular combination of *Hox* proteins, and there is a specific blueprint (*Hox* code) that instructs each cell as to which part of the body it properly belongs (Kessel and Gruss, 1991; Jeannotte et al., 1993).

While there is much research to be done, the general conclusion that is emerging from these various studies is that the main body axis of an animal, the front-to-rear (anteroposterior) organization of most metazoans, is specified by arrays of homeobox genes that are remarkably similar in nematode worms, fruit flies, and mammals, including humans. This, in turn, has had substantial impact on our views of the evolution of the Metazoa.

derepress

How Do *Hox* Genes Work?

Very little is known about the detailed molecular mechanisms by which homeobox genes affect developmental patterns, whether in *Drosophila* or in vertebrates. Although it is clear that *Hox* DNA-binding proteins act by turning on a large number of subordinate, **target genes**, the specific targets are as yet mainly unknown.

One interesting finding is that homeodomain-containing proteins can regulate the activity of target genes that produce the sticky molecules on cell surfaces that hold them together to make up a multicelled tissue. For example, in mammalian cells grown in the laboratory, the *Hoxb 9* and *Hoxb 8* proteins have been shown to regulate the chemical switch (the region of the chromosome known as the **promoter**) that turns on the manufacture of the Neural-Cell Adhesion Molecule (N-CAM) (Jones et al., 1992). Similarly, the genes targeted by *Ultrabithorax* include *connectin*, a gene that produces a molecule that causes cells to stick together and is involved in establishment of nerve connections in muscle tissue (Gould and White, 1992), and *scabrous*, a gene involved in intercellular communication during the formation of nerve tissue (Graba et al., 1992).

Another interesting finding is that in nematode worms, an *Antp*-like gene acts as a switch that controls migration of particular cells that, during the development of the worm, give rise to its nervous system (Salser and Kenyon, 1992). This relationship between homeobox genes and cell movement is also found in the frog homeobox gene called *goosecoid*, which directs cell migration during a very early (gastrulation) stage of embryonic development (Niehrs et al., 1993). Thus, the homeodomain proteins may affect pattern formation by regulating cell adhesion and migration, but this needs further study.

When homeobox genes first evolved, their original function may have been to control cell migration. According to this scenario, the motile single-celled ancestor of metazoans would have had a set of genes that contolled its movements, **regulatory genes** that functioned much like the gearbox of an automobile, regulating the cell's movement toward food or in response to other signals detected in the local environment. Over time, such cells aggregated, ultimately giving rise to multicelled metazoans, and these regulatory genes, already in place, could have been co-opted during subsequent evolution to provide information instructing individual cells to occupy particular positions within the growing organism. Although there is no evidence yet for the presence of homeobox complexes in unicellular forms of life, this possibility should be investigated more fully. If homeotic complexes are absent from all single-celled organisms, it may mean that they first appeared in the earliest metazoans and that the origin of these gene complexes was thus one of the most fundamental of all steps in the evolutionary development of the basic body plan of animals.

EVOLUTION OF BODY FORM

Homeobox genes have been extraordinarily conserved throughout all of animal evolution. From hydra to nematode worms, from fruit flies to mammals, homeobox genes seem to orchestrate pattern formation, the overall development of body form in animals.

With four *Hox* complexes in mammalian chromosomes, evolution has been provided a fertile source for the derivation of a large variety of vertebrate body shapes. Changes in a single *Hox* gene altered the vertebrae of mice (Figure 5.10), a homeotic transformation evidently quite similar to those found in sloths (Table 5.1) and even humans (Figure 5.3). However, perhaps the most striking example of how *Hox* genes may have played a role in the evolution of the vertebrate body is provided by a recent experiment by Lufkin and his coworkers (1992). They altered the promoter (the chomosomal region that directs whether a gene is active) of a *Hox* gene in the anterior end of mice so that it was artificially much more active than normal. Lufkin's group observed a remarkable transformation, namely, one of the mouse head bones, the occipital, was converted into four vertebrae. This result is startling and especially interesting because more than 500 million years ago, during the Cambrian Period of Earth history when the vertebrate head was first evolving, head bones at the base of the skull were derived from four ancestral vertebrae! Primitive jawless fishes, such as the lampreys of the present day, have four extra vertebrae. And all animal embryos, including those of humans, have four anterior body segments (somites) that instead of forming vertebrae are incorporated into the occipital bone of the head. Thus, artificial stimulation of a single *Hox* gene in a living mouse was sufficient to produce bone structures that had disappeared hundreds of millions of years ago. Clearly (as discussed also in Chapter 4), modern vertebrates contain a genetic memory of the earliest stages of their evolutionary development.

New species of multicellular animals arise not through changes in their biochemical capabilities but, rather, through changes in how the cells of their bodies are arranged with respect to each other. The position of cells in all metazoan organisms appears to be specified by a single universal set of molecular mechanisms. The range of possible changes in evolution is constrained by those developmental mechanisms that specify the way in which cells build the animal body. Mutations in the master regulatory genes that guide the development of embryos, such as *Hox* genes and others, may have played an important role in providing the variation in form and structure on which natural selection, the keystone of the evolutionary process, has acted over time. The study of homeotic genes has produced significant progress in our understanding of the molecular mechanisms of organismal development, and this, in turn, has energized studies of the evolution of one of the most fundamental of all metazoan characteristics, the basic body plan of animals.

■

REFERENCES

Awgulewitsch, A., and Jacobs, D. 1990. Differential expression of Hox 3.1 protein in subregions of the embryonic and adult spinal cord. *Development 108:* 411–420.

Bateson, W. 1894. *Materials for the Study of Variation Treated with Especial Regard to Discontinuity in the Origin of Species* (London: Macmillan).

Boncinelli, E., Somma, R., Acampora, D., Pannese, M., D'Esposito, M., Faiella, A., and Simeone, A. 1988. Organization of human homeobox genes. *Hum. Reprod. 3:* 880–886.

Bridges, C.B., and Morgan, T.H. 1923. *The Third-Chromosome Group of Mutant Characters of Drosophila melanogaster* (Washington: Carnegie Institution), pp. 137–139.

Bürglin, T.R., Ruvkun, G., Coulson, A., Hawkins, N.C., McGhee, J.D., Schaller, D., Wittman, C., Müller, F., and Waterston, R.H. 1991. Nematode homeobox cluster. *Nature 351:* 703.

Carrasco, A.E., McGinnis, W., Gehring, W.J., and De Robertis, E.M. 1984. Cloning of an *X. laevis* gene expressed during early embryogenesis coding for a peptide region homologous to *Drosophila* homeotic genes. *Cell 37:* 409–414.

De Robertis, E.M., Wright, C.V.E., and Oliver, G. 1990. Homeobox genes and the vertebrate body plan. *Scientific American 263:* 42–55.

Duboule, D., and Dollé, P. 1989. The structural and functional organization of the murine HOX gene family resembles that of *Drosophila* homeotic genes. *EMBO J. 8:* 1497–1505.

Gaunt, S.J., Sharpe, P.T., and Duboule, D. 1988. Spatially restricted domains of homeo-gene transcripts in mouse embryos: Relation to a segmented body plan. *Development Suppl. 104:* 169–179.

Gehring, W. 1994. In: D. Duboule (Ed.), *Guidebook of Homeobox Genes* (Oxford: IRL Press), pp. 1–10.

Gould, A.O., and White, R.A.H. 1992. *Connectin*, a target of homeotic gene control in *Drosophila*. *Development 116:* 1163–1174.

Graba, Y., Aragnol, D., Laurenti, P., Garzino, V., Charmot, D., Berenger, H., and Pradel, J. 1992. Homeotic control in *Drosophila*; the *scabrous* gene is an *in vivo* target of *Ultrabithorax* proteins. *EMBO J. 11:* 3375–3385.

Graham, A., Papalopulu, N., and Krumlauf, R. 1989. The murine and *Drosophila* homeobox gene complexes have common features of organization and expression. *Cell 57:* 367–378.

Jeannotte, L., Lemieux, M., Charron, J., Poirier, F., and Robertson, E.J. 1993. Specification of axial identity in the mouse: Role of the *Hoxa-5 (Hox1.3)* gene. *Genes Dev. 7:* 2085–2096.

Jegalian, B.C., and De Robertis, E.M. 1992. Homeotic transformations in the mouse induced by overexpression of a human *Hox 3.3* transgene. *Cell 71:* 901–910.

Jones, F.S., Prediger, E.A., Bittner, D.A., De Robertis, E.M., and Edelman, G.M. 1992. Cell adhesion molecules as targets for Hox genes: N-CAM promoter activity is modulated by cotransfection with Hox 2.5 and 2.4. *Proc. Natl. Acad. Sci. USA 89:* 2086–2090.

Kenyon, C., and Wang, B. 1991. A cluster of *Antennapedia*-class homeobox genes in a nonsegmented animal. *Science 253:* 516–517.

Kessel, M., and Gruss, P. 1991. Homeotic transformations of murine vertebrate and concomitant alteration of *Hox* codes induced by retinoic acid. *Cell 67:* 89–104.

Kühne, K. 1932. Die Vererbung der Variationen der Menschlichen Wirbelsäule. *Z. Morph. Anthropol. 30:* 1–221.

Kühne, K. 1936. Die Zwillingswirbelsäule. *Z. Morph. Anthropol. 35:* 1–376.

Le Mouellic, H., Lallemand, Y., and Brûlet, P. 1992. Homeosis in the mouse induced by a null mutation in the *Hox-3.1* gene. *Cell 69:* 251–264.

Lewis, E.B. 1978. A gene complex controlling segmentation in *Drosophila*. *Nature 276:* 565–570.

Lufkin, T., Mark, M., Hart, C. Doll, P., LeMeur, M., and Chambon, P. 1992. Homeotic transformations of the occipital bones of the skull by ectopic expression of a homeobox gene. *Nature 359:* 835–841.

Malicki, J., Schughart, K., and McGinnis, W. 1990. Mouse *Hox-2.2* specifies thoracic segmental identity in *Drosophila* embryos and larvae. *Cell 63:* 961–967.

McGinnis, W., Levine, M., Hafen, E., Kuroiwa, A., and Gehring, W.J. 1984. A conserved DNA sequence in homeotic genes of the *Drosophila Antennapedia* and *bithorax* complexes. *Nature 308:* 428–433.

McGinnis, N., Kuziora, M.A., and McGinnis, W. 1990. Human *Hox*-4.2 and *Drosophila deformed* encode similar regulatory specificities in *Drosophila* embryos and larvae. *Cell 63:* 969–976.

McGinnis, W., and Krumlauf, R. 1992. Homeobox genes and axial patterning. *Cell 68:* 283–302.

Murtha, M.T., Leckman, J.F., and Ruddle, F.H. 1991. Detection of homeobox genes in development and evolution. *Proc. Natl. Acad. Sci. USA 88:* 10711–10715.

Niehrs, C., Keller, R., Cho, K.W.Y., and De Robertis, E. 1993. The homeobox gene *goosecoid* controls cell migration in *Xenopus* embryos. *Cell 72:* 491–503.

Pollock, R.A., Jay, G., and Bieberich, C.J. 1992. Altering the boundaries of *Hox 3.1* expression: Evidence for antipodal gene regulation. *Cell 71:* 911–923.

Rabin, M., Hart, C.P., Ferguson-Smith, A., McGinnis, W., Levine, M., and Ruddle, F.H. 1985. Two homeobox loci mapped in evolutionary related mouse and human chromosomes. *Nature 314:* 175.

Raff, R.A., and Kaufman, T.C. 1983. *Embryos, Genes and Evolution* (New York: Macmillan).

Salser, S.J., and Kenyon, C. 1992. Activation of a *C. elegans Antennapedia* homologue in migrating cells controls their direction of migration. *Nature 355:* 255–258.

Schierwater, B., Murtha, M., Dick, M., Ruddle, F.H., and Buss, L.W. 1991. Homeoboxes in Cnidarians. *J. Exp. Zool. 260:* 413–416.

Schummer, M., Scheurlen, I., Schaller, C., and Galliot, B. 1992. HOM/HOX homeobox genes are present in hydra *(Chlorohydra viridissima)* and are differentially expressed during regeneration. *EMBO J. 11:* 1815–1823.

Scott, M.P., and Weiner, A.J. 1984. Structural relationships among genes that control development: Sequence homology between the *Antennapedia*, *Ultrabithorax* and *Fushi Tarazu* loci of *Drosophila*. *Proc. Nat. Acad. Sci. USA 81:* 4115–4119.

Scott, M.P. 1992. Vertebrate homeobox gene nomenclature. *Cell 71:* 552–553.

FROM MOLECULAR EVOLUTION TO BIOMEDICAL RESEARCH: THE CASE OF CHARLES DARWIN AND CHAGAS' DISEASE

■

Larry Simpson*

DARWIN'S ILLNESS

Charles Darwin (Figure 6.1) has always presented a fascination for everyone interested in the history of science, especially those interested in development of the theory of evolution. Here was a man who single-handedly changed the conceptual framework of all biological science and the way we look at the living world around us, but who led a quiet, outwardly uneventful life. In fact, the one aspect of Darwin's personal life that was unusual was his state of health. For over half his life, Darwin suffered from an undiagnosed chronic illness that severely affected both his ability to do science and his interactions with people in general. His symptoms, which included extreme fatigue and digestive and intestinal problems, were frequently almost incapacitating. That Darwin was able to produce such an impressive body of work is truly amazing.

For the first 27 years of his life, until 1838, Darwin was in excellent health. In the summer of 1826, Darwin and two friends took a walking tour through North Wales, often covering 30 miles a day and climbing local peaks (Adler, 1989). At Cambridge University, Darwin was a member of the Gourmet Club and his diary includes many entries about eating and drinking (Adler, 1989). As the ship's naturalist on the *H.M.S. Beagle* during a voyage that lasted from 1831 to 1836 (Figure 6.2), Darwin was considered one of the most fit members of the crew and he frequently went ashore and undertook long overland expeditions (Adler, 1989). His only illness during this trip (aside from continual sea sickness) was a month-long bout of fever of unknown origin in 1834.

In 1836, Darwin returned to England and was in good health for the next two years until his chronic symptoms began to appear and his life changed permanently. In 1838, he refused the secretaryship of the Geological Society of London for reasons of ill health,

*Howard Hughes Medical Institute, Department of Biology, and Department of Medical Microbiology and Immunology, University of California, Los Angeles, CA 90095.

FIGURE 6.1

Charles Darwin in 1840 at age 31, shown in a watercolor portrait by George Richmond. (Reproduced, with permission, from Moorehead, 1969.)

and was advised by his doctors to stop working temporarily (Goldstein, 1989). Darwin's health became progressively worse during the next few years. He suffered from periodic vomiting and debilitating fatigue. In 1839, he married, and in 1842, his family moved to a country home, thinking that the peaceful surroundings would improve his health. However, he wrote to botanist Joseph Hooker in 1845:

> I believe that I have not had not a whole day, or rather night, without my stomach being greatly disordered, during these last three years, and most days great prostration of strength.

Darwin also recorded the fact that after social dinners with his friends he commonly suffered "violent shivering and vomiting attacks" and that because of this unpleasantness he was "compelled for many years to give up all dinner parties." In 1849, he was too ill to attend his father's funeral. He wrote: "I was almost quite broken down, head swimmy, hands trembling and never a week without violent vomiting."

Throughout his life, Darwin was examined by many eminent physicians who could find no organic cause of his symptoms. During his last ten years, from 1862–1872, Darwin's health improved somewhat, but his symptoms never fully disappeared.

Many theories have been proposed to account for Darwin's mysterious illness. These range from suggestions that he was a hypochondriac, that he suffered an Oedipal complex or a psychosis, that he had an allergy to pigeons—even that he was being poisoned by the arsenic in his medications—to the idea that he had an undiagnosed chronic disease such as brucellosis, a bacterial infection (Goldstein, 1989).

CHAGAS' DISEASE

In a compelling paper published in 1959, Saul Adler, an Israeli parasitologist, suggested that Darwin suffered from chronic Chagas' disease and that he contracted this disease during the voyage of the *Beagle* (Adler, 1959). Chagas' disease is an incurable chronic disease caused by the **eukaryotic** (cell nucleus-containing) protozoal parasite, *Trypanosoma cruzi*. It affects the heart, the involuntary (**autonomic**) nervous system, or both, and is a major cause of sudden death in Latin America. Approximately 16 to 18 million people inhabiting the region from Argentina to Mexico are infected with this parasite.

The *T. cruzi* parasite belongs to a large group of flagellated protozoa known as **kinetoplastid** protozoa, named for the presence of a large mass of circularly organized DNA (kinetoplast DNA) that occurs in the energy-producing organelle (**mitochondrion**) at the base of the whiplike **flagellum** that enables these protozoans to swim (Simpson, 1987; Simpson, 1972) (Figures 6.3 and 6.4). This DNA consists of a giant network of thousands of interlocked small circles, **catenated minicircles**, and a smaller number of larger catenated **maxicircles** (Figures 6.5 and 6.6). This may be one of the most unusual DNA structures known in nature; it has been the subject of intensive research in several laboratories, including my own.

Chagas' disease was discovered in 1909 by the Brazilian naturalist, Carlos Chagas, who found the parasites in insects and other animals and decided to look for them in sick humans (Figure 6.7). Chagas completely described both the life cycle of the parasite in the **insect vector** (the host organism of the parasite that transmits the parasite to vertebrates at one stage in its life cycle) and in the human or animal host. He also described in detail the epidemiology of the disease and the pathology it caused (see Prata, 1981, for a collection of Chagas' papers). Chagas' disease is one of only a few examples of a human disease that was discovered *after* the disease-causing parasite had been found in nature. Chagas named the parasitic species *T. cruzi*, in honor of Oswaldo Cruz, another distinguished Brazilian scientist. Chagas' work led to the establishment of a Brazilian tradition

FIGURE 6.2

Map of a portion of the Beagle's voyage, showing Darwin's land excursions in South America. (Reproduced, with permission, from Stone, 1980.)

THE BEAGLE VOYAGE

of research into the biology, biochemistry, and, recently, the molecular biology, of these widespread parasites.

The parasite is transmitted by **reduviid bugs**, a particular type of insect belonging to the genera *Triatoma, Rhodnius,* and *Panstrongylus.* These insects live in the walls and thatched roofs of adobe huts, and they emerge at night to feed on sleeping people (Figure 6.3C). In Brazil, the insects are known as "kissing bugs," because they frequently bite the face of their victim near the mouth or eyes, or as "barber bugs," because they engorge themselves with blood like the leeches used by ancient barbers. Their bloodsucking has also caused them to be labeled "assassin bugs." The *T. cruzi* parasites living in the reduviids are excreted in bugs' feces, which they deposit soon after a blood meal; a sleepy person who has been bitten may rub the parasite-containing fecal material into the wound and thereby initiate the infection.

Some term Chagas' disease a *social disease*, because better living conditions eliminate the insect vector and stop the normal mode of transmission. However, even with improved living conditions, transmission by blood transfusions or from a mother to her unborn fe-

FIGURE 6.3

Trypanosoma cruzi and its insect vector. Parts A and B: Light photomicrographs showing *T. cruzi* in a stained blood film from an infected person; note the intensely stained kinetoplast (k) at the base of the flagellum (f). Part C: A reduviid bug taking a blood meal; reduviids in the genera *Triatoma, Rhodnius,* and *Panstrongylus* transmit *T. cruzi* by fecal contamination. (Reproduced, with permission, from Peters and Gilles, 1977.)

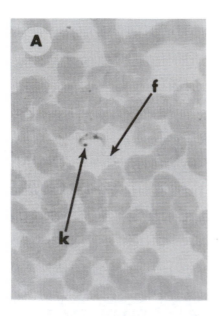
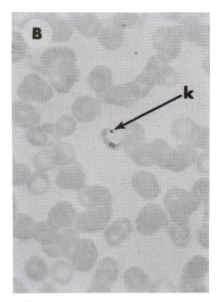
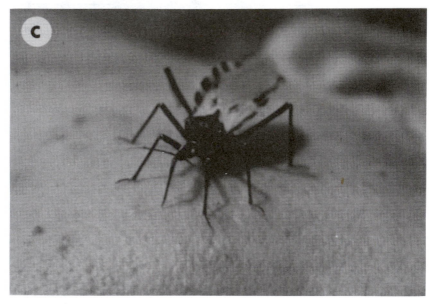

tus would still be major problems. In fact, transmission by transfusion has recently become a problem in the Los Angeles Blood Bank because of donation of blood by infected people who have immigrated from Central America (Kirchoff, 1989; Kirchoff et al., 1987).

Once inside the host, the *T. cruzi* parasites migrate rapidly through the blood to cells of heart muscles or of the autonomic nervous system, which they penetrate and in which they survive and live as intracellular parasites. The acute phase of Chagas' disease usually occurs in children and involves fever, swelling of the lymph nodes, spleen, and liver, and serious damage to the heart (myocarditis). Some mortality occurs in the acute phase, but most people survive and the symptoms disappear after several months. Approximately 20% to 40% of infected people develop the chronic form of the disease (Schmunis, 1991), which involves progressive degeneration of the heart, of the gastrointestinal system, or of both. After infection, there is usually a lag period of three to five years before the onset of the chronic symptoms, which last for the rest of the life of the diseased victim.

The "Benchuca" Incident on the Voyage of the *Beagle*

Adler (1959) noted that many of the symptoms Darwin reported matched those of chronic Chagas' disease. Moreover, Adler found written on March 26, 1835, in Darwin's chronicle of the expedition, *The Voyage of the Beagle*, a passage describing an overland trip Darwin took from Chile across the Andes Mountains to the plains of Argentina (Figures 6.2 and 6.8):

> We slept in the village, which is a small place, surrounded by gardens, and forms the most southern part, that is cultivated, of the province of Mendoza; it is five leagues south of the capital. At night I experienced an attack (for it deserves no less a name) of the Benchuca (a species of *Reduvius*) the great black bug of the Pampas. It is most disgusting to feel soft, wingless insects, about an inch long, crawling over one's body. Before sucking, they are quite thin, but afterwards be-

---

FIGURE 6.4

Electron micrographs of kinetoplast DNA in *Trypanosoma cruzi* cells. Parts A to C: Longitudinal sections through the kinetoplast-containing (K) portion of the single mitochondrion, showing the kDNA (DNA), mitochondrial tubules (M), nucleus (N), flagellum (Fl), nucleolus (Nu), and Golgi (G); the white arrow in the lower right portion of Part B points to an amorphous mass of kDNA at one end of the nucleoid body, which may represent a site for minicircle DNA replication. Part D: Transverse section through the kinetoplast nucleoid body showing the minicircular DNA fibrils. (Reproduced, with permission, from Delain and Riou, 1969.)

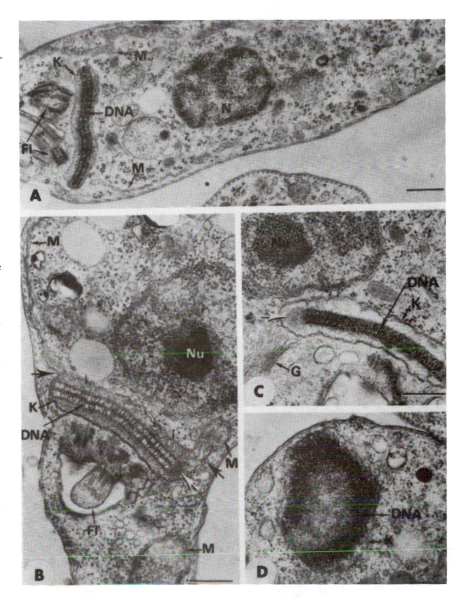

FIGURE 6.5

Electron micrograph of a kinetoplast DNA network from *Leishmania tarentolae*. The DNA was dispersed on water, picked up on a carbon surface, and shadowed with platinum/paladium. The interlocked (catenated) minicircle and maxicircle DNA molecules are well defined (but are especially evident at higher magnification, as shown in Figure 6.6). (Reproduced, with permission, from Simpson and Berliner, 1974.)

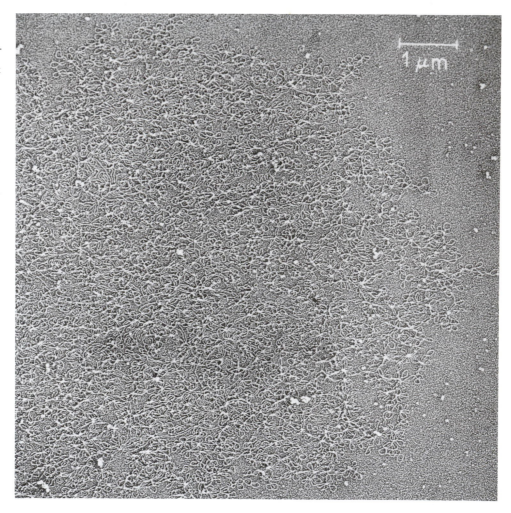

come round and bloated with blood, and in this state they are easily crushed. They are also found in the northern parts of Chile and in Peru.

The "great black bug of the Pampas" was probably the reduviid *Triatoma infestans*—more than 70% of specimens of that species in that region of Argentina are infected with the *T. cruzi* parasite at the present time (Schmunis, 1991). In addition, approximately 12% of the human population today living in that region have been shown to contain **antibodies** (biochemicals of the immune system) against *T. cruzi* (Schmunis, 1991). This passage provides one specific instance in which Darwin could have been infected with *T. cruzi*. Furthermore, the three- to five-year period until the onset of his symptoms agrees with the known lag period for the onset of chronic Chagas' disease.

Diagnosis of Chagas' Disease

Adler's theory has created much controversy and has not been accepted by everyone, mainly because it has been impossible to verify (Adler, 1989; Bernstein, 1984; Goldstein, 1989). Chagas' disease was undescribed until the early 1900s, more than 70 years after Darwin visited the Argentinian Pampas, and even today it is difficult to diagnose in chronic patients because the parasites are normally localized within cells of heart and nerve tissue; only very few parasites circulate in the blood. Remarkably, the best method for clinical diagnosis of the presence of these parasites in a patient is **xenodiagnosis**, in which uninfected kissing bugs are actually allowed to feed on a patient (!), are grown in the laboratory for several weeks, and are then dissected to see whether parasites are present in their hind guts (Figure 6.9). Although this method (aside from having obvious esthetic problems) lacks sensitivity and reproducibility, it nevertheless is the current gold standard for detection of parasites in a patient. Detection of antibodies against the parasite is also used, but the reagents employed frequently react with other antibodies as well, and therefore the test is not entirely reliable.

KINETOPLAST DNA: MINICIRCLES AND MAXICIRCLES

This laboratory has been involved in studies of the genetic material (the "genome") of the kinetoplast of the mitochondrion (see Box 6.1) in kinetoplastid protozoa since 1969. In 1980, a collaboration began with Carlos Morel from the Oswaldo Cruz Institute in Rio de Janeiro on the kinetoplast DNA (**kDNA**) of *T. cruzi*. From 1917 to 1934 Carlos Chagas, the discoverer of Chagas' disease, directed the Oswaldo Cruz Institute (Figure 6.10), and Morel is currently President of the Oswaldo Cruz Foundation that administers the In-

FIGURE 6.6

Electron micrograph of a fragment of a kinetoplast DNA network from *Leishmania tarentolae*, showing catenated minicircles and "figure of eight" DNA molecules (which may be minicircle recombinants). The precise molecular topology within the network is not understood. (Previously unpublished electron micrograph provided by Dr. Agda Simpson, University of California, Los Angeles.)

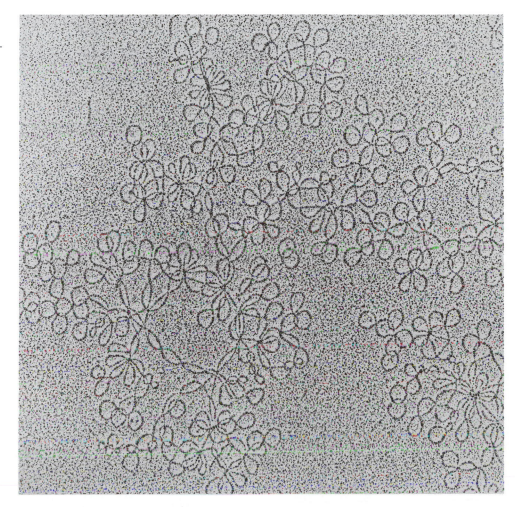

Carlos Chagas (center) and members of an expedition from the Oswaldo Cruz Foundation on the Rio Negro River in Amazonas, Brazil, 1913. (Reproduced, with permission, from Thielen, 1991.)

stitute. Morel and I found that the kDNA from different strains of *T. cruzi* differed greatly in the sequence of the subunits of which they are composed (Morel et al., 1980). This was shown by digesting the kDNA with special enzymes (**restriction enzymes**) that cleave DNA at specific places in the molecule and, as illustrated in Figure 6.11, by separating the cleaved fragments with use of high resolution **gel electrophoresis** (a technique in which the cleaved segments are subjected to an electrical charge that pulls them across a gel; because short pieces move faster than longer ones, the fragments are separated into a pattern of bands that will be identical from one sample to another only if the kDNAs are the same). Because the minicircle DNA makes up 95% of the total kDNA network, we believed that the sequence differences detected were caused by the DNA in the mini-circles. We coined the term, **schizodeme**, to refer to various strains that distinctly differ one from another in their kDNA minicircle sequences (Morel et al., 1980). The genetic reason for the large amount of sequence divergence was shown only recently in genetic studies by Tibayrenc, Ayala, and colleagues to be a result of the fact that the various *T. cruzi* schizodemes actually represent major evolutionary lineages, genetically isolated from each other for such long periods of time that the minicircle sequences have diverged ex-tensively (Tibayrenc et al., 1986; Tibayrenc et al., 1990).

In the early 1980s, we did not know the genetic function of the minicircle molecule, which is present in high abundance in the kinetoplast. The other molecular species in the network, known as maxicircle DNA, appeared to represent the evolutionarily related **ho-**

FIGURE 6.8

A long, narrow basin separating the Uspallata Range from the Andes mountain chain, north of Mendoza, Argentina, near where Darwin stayed overnight and was bitten by a "Benchuca" bug. The curious structure in the distance probably represents an uplifted rock unit. (Reproduced, with permission, from Moorehead, 1969, after Schmidtmeyer, 1824.)

mologue of the informational DNA molecule found in human mitochondria (Simpson, 1987).

RNA EDITING

The DNA strands that contain genetic information in cells are each made up of nearly three billion subunits, called **nucleotides**, each of them representing one of the four "letters" in an information-containing chemical code (A, adenine; T, thymine; C, cytosine; and G, guanine). One of the basic tenets of molecular genetics is that the nucleotide sequence in the DNA should be copied faithfully, as determined by rules of chemical bonding (the ability to make base pairs, discussed in Chapter 2), into a molecule known as **messenger RNA (mRNA)**—the molecule that carries the coded message from the DNA to ribosomes in which proteins are manufactured. The mRNA copies should be perfect, except for lacking intervening sequences (stretches equivalent to the "junk" DNA that is

FIGURE 6.9

Xenodiagnosis. Uninfected reduviid bugs are permitted to take a blood meal on the patient, and are then grown in the laboratory for several weeks and examined for the presence of *Trypanosoma cruzi* in their hind guts. (Reproduced, with permission, from Peters and Gilles, 1977.)

interspersed between information-containing segments in the genome) and having an added 3′ **poly[A] tail** and an added 5′ **cap structure** at their ends. In the late 1980s it was discovered that, in addition to normal genes, the maxicircle DNA molecule in the mitochondrion of kinetoplastid protozoa contains hidden genes. The RNA strands copied from these **cryptogenes** are modified (that is, they are **edited**) by the insertion and deletion of **uridine** (U) residues, and the modified, edited message can then be translated at ribosomes to produce protein products (Benne et al., 1986; Feagin et al., 1988; Shaw et al., 1988; Simpson et al., 1993; Stuart, 1993). However, there appeared to be no DNA source for the new information represented by the addition and deletion of these uridine residues. The information for the modified nucleotide sequence seemed to come from nowhere! These surprising results led some scientists to question the well-entrenched **central dogma** of molecular genetics—that genetic information flows from DNA to mRNA and from mRNA to protein—because the genetic information for the sequence changes in the mRNAs (information that determines the sequence of amino acids in the proteins they produce, discussed in Chapter 2) did not appear to reside in the DNA of the organism. We now believe that these results can be explained in terms of a model that is consistent with the central dogma.

Central to this model is the existence of a new class of RNA molecules, known as **guide RNAs (gRNAs)** (discussed in Box 6.2), that contain short pieces made up of the information required to produce the edited nucleotide sequences. We discovered gRNAs in 1989 by performing a computer search of the known maxicircle sequence, looking for short stretches complementary to (that is, stretches that could base pair with) the edited sequences. We found several that fit the model if we allowed G (guanine) residues to base pair with U (uridine) residues in addition to the C (cytosine) residues with which they more commonly pair (Blum et al., 1989). We then showed that the predicted gRNAs actually exist in *T. cruzi*, and that they consist of short RNA molecules, approximately 50 nucleotides in length, that have a nonencoded stretch of U residues at one end. The gRNAs form **duplex "anchors"** by hybridizing with (bonding to) specific mRNA sequences adjacent to the region to be edited and that contain additional A (adenine) and G residues, both of which can form base pairs with the U residues inserted by editing. Fully edited mRNA makes a perfect duplex with the gRNA. Thus, our newly discovered guide

BOX 6.1

Mitochondria.

Mitochondria are the organelles in eukaryotic cells that produce chemical energy, stored in molecules of ATP (adenosine triphosphate), from the stepwise breakdown (oxidation) of foodstuffs such as glucose. They have two membranes, the inner one of which contains particular (cytochrome and flavoprotein) enzymes that make up the electron transport chain involved in energy production and the enzymes needed for synthesis of ATP. Within the matrix of the mitochondrion are located the enzymes of the TCA (tricarboxylic acid) cycle that converts pyruvate to acetyl Coenzyme-A and in so doing releases electrons that enter the cytochrome chain of the inner membrane. The matrix also contains the mitochondrial DNA, and mitochondrial ribosomes and associated translation machinery in which a small number of proteins is manufactured by using the information encoded in the DNA. The amount of genetic information encoded in animal and fungal mitochondrial DNA is variable, but is usually limited to that required for synthesis of cytochrome *b* and of several subunits each of cytochrome oxidase, NADH dehydrogenase, and ATP. Two ribosomal RNAs and a set of 24 transfer RNAs are also usually transcribed from the mitochondrial DNA, which in animal and fungal cells contains a sequence of information-containing nucleotides about 15 kb (kilobases) long. In higher plants, the amount of DNA in the mitochondria is much larger and their mitochondrial genomes contain many more genes.

FIGURE 6.10

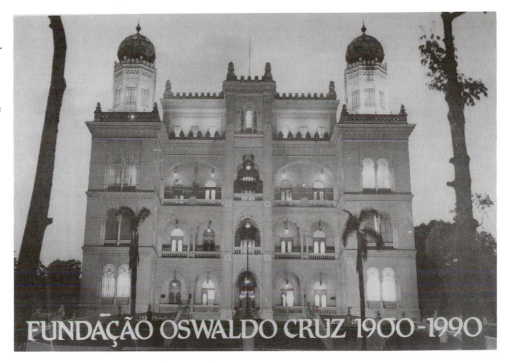

Oswaldo Cruz Institute in Rio de Janeiro, Brazil, shown in a poster celebrating the first 90 years of the Fundacao Oswaldo Cruz. This "little castle," the oldest building in the Institute, is currently used for administration and the library.

FIGURE 6.11

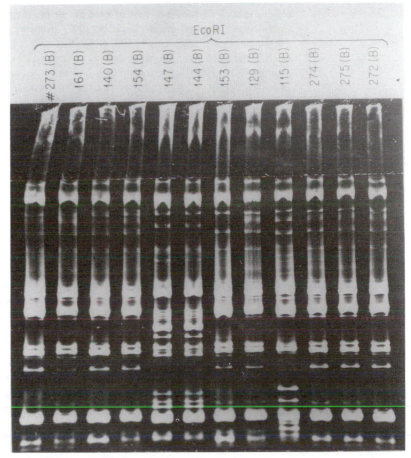

Acrylamide gel electrophoresis of kinetoplast DNA from *Trypanosoma cruzi* isolated from several human patients. The *T. cruzi* parasites were isolated from human blood and grown in laboratory cultures, and the kDNA was isolated and digested with the EcoRI restriction enzyme and electrophoresed on an acrylamide gradient gel. The bands, which are made visible by staining with ethidium bromide, mainly represent fragments of minicircle DNA. The numbers refer to individual patients, all of whom share a common pattern of isoenzymes (B). Note the two distinct types of banding patterns, one group consisting of patients 147, 144, and 115, and the other group consisting of the remaining patients. (Reproduced, with permission, from Morel et al., 1980.)

BOX 6.2   **Guide RNAs in flagellated protozoa.**

The flagellated protozoa known as kinetoplastids contain a single mitochondrion that has a huge mass of mitochondrial DNA situated adjacent to the base of the flagellum. In trypanosomatids, a subgroup of kinetoplastids, this DNA consists of thousands of catenated minicircles and a smaller number of maxicircles all linked into a giant DNA network. The maxicircles are the evolutionarily related homologues of the mitochondrial DNA of human cells and contain the information for synthesis of two ribosomal ribonucleic acids (rRNAs) and several types of protein. The function of the minicircles was not understood until recently. Benne et al. (1986) and Shaw et al. (1988) discovered that several transcripts of the maxicircles are modified after transcription by the insertion and deletion of uridine residues within coding regions at multiple sites to form an "open reading frame" that encodes a protein. The extent of this RNA editing varies from insertion of four Us at three sites (which therefore offsets a "−1" frameshift) to insertion of hundreds of Us at hundreds of sites, creating an open reading frame apparently "from whole cloth" (Simpson et al., 1993). The information for this editing is contained in small RNAs, called guide RNAs (gRNAs) because they guide the editing process, which are transcribed both from maxicircle and from minicircle DNA molecules (Simpson et al., 1993). The 5′ ends of the gRNAs can form duplex "anchor" regions just downstream from the region to be edited (the region known as the PER), and the remainder of the molecule forms perfect duplexes with fully edited RNAs.

Two basic models have been proposed for the mechanism of the editing. The enzyme cascade model (Blum et al., 1990) proposes that an endonucleolytic cleavage occurs at the site of the first mismatch between the mRNA and the gRNA, the addition of a U to the 3′ hydroxyl group at the cleavage site, and a ligation (a bonding together) of the two resulting mRNA fragments. The added U bonds, by base pairing, with a guide A or G residue in the gRNA, and thereby extends the anchor duplex. The other model, the transesterification model (Blum and Simpson, 1992), proposes that the Us are transferred from the nonencoded 3′ oligo-U tail of the gRNA by two successive chemical reactions (transesterifications) that are like those that occur during RNA splicing. Whatever the precise chemistry, it is clear that a single gRNA molecule mediates the editing of a single block of nucleic acid, and that multiple overlapping gRNAs mediate the editing of multiple contiguous blocks (a "domain" of nucleic acid), as shown in Figure 6.12 for the MURF4 gene of the parasite *L. tarentolae*. In this lizard parasite, there are a limited number of gRNAs, each of which encodes a specific editing block. In contrast, in *Trypanosoma brucei* there are more than 1000 gRNAs that show extensive overlap even of their coding regions (Corell et al., 1993). The reason for this difference in the editing mechanism of the gRNAs in these two species of parasites is not clear, but it may be a result of changes that occurred during their evolutionary histories.

RNAs had come to the rescue of the central dogma. (We were pleased but also somewhat chagrined that the answer to the secret of editing was not something completely unexpected but was instead a novel system that still obeyed the simple rules of base pairing.) As discussed in Box 6.2, the precise mechanism by which the gRNAs mediate the insertion of U residues is still uncertain.

We then discovered that the genetic role of the previously mysterious minicircle kDNA molecules was to encode gRNAs, that is, the kDNAs contain the genetic information that results in formation of gRNA molecules (Sturm and Simpson, 1990; Pollard et al., 1990). A few gRNAs were encoded by the maxicircle DNA, but most were found to be encoded by the minicircles. In the lizard parasite *Leishmania tarentolae*, a trypanosomal parasite related to *T. cruzi*, there is a single gRNA gene per minicircle, whereas in *T. cruzi* there are several gRNAs encoded in each minicircle (Avila and Simpson, unpublished results).

Editing (the insertion or deletion of uridine residues) was found to vary in different genes and in different kinetoplastid protozoal species from having a few Us inserted at a few sites to having hundreds of Us inserted at hundreds of sites. This latter, very extensive editing, termed **pan-editing**, involves multiple gRNAs that act sequentially in a complex process in which the first gRNA creates the anchor sequence for the second gRNA, the second makes the anchor for the third, and so forth (Maslov and Simpson, 1992). An example of a pan-edited cryptogene, together with the relevant overlapping gRNAs from *L. tarentolae*, is shown in Figure 6.12.

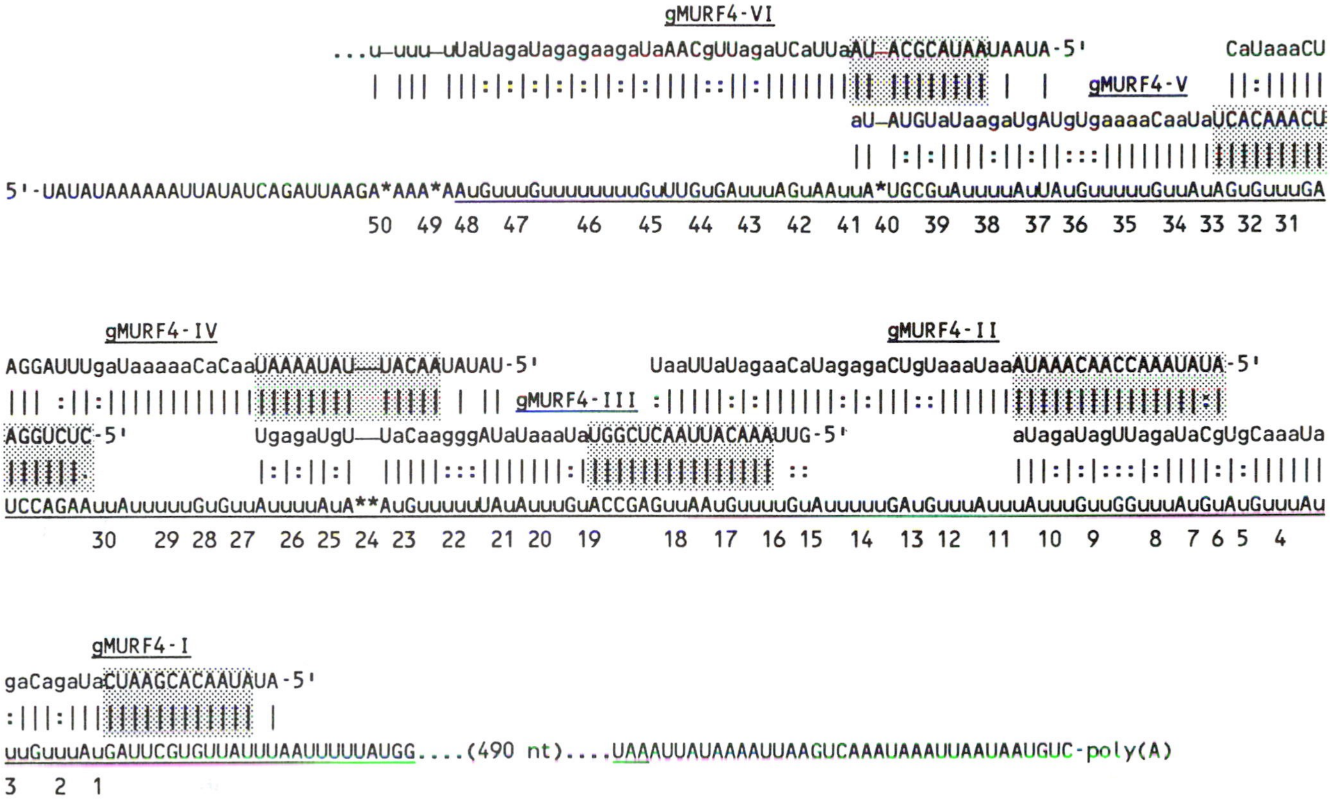

FIGURE 6.12

Pan-editing of transcripts from a maxicircle cryptogene of *Leishmania tarentolae*. The nucleotide sequence of the edited mRNA that encodes the information for protein synthesis (subdivided into three rows in the figure) is identified by the underlining. The uridines (u) inserted by editing increase the length of the original sequence encoded in the unedited gene by about twofold. The initial mRNA complimentary to the DNA of the gene is modified at 50 locations by RNA editing, as indicated by the numbers. The overlapping gRNAs (for example, "gMURF4-I"), which mediate the editing process, are shown above the sequence. Editing begins at gMURF4-I (lower left) and ends with gMURF4-VI. The gRNA/mRNA "anchor" duplex region is shaded for each editing block. The uridines inserted by editing are indicated by (u), and the guide nucleotides in the gRNAs by (a) or (g). Encoded (u) edited out of the mRNA are indicated by asterisks. G-C and A-U base pairs are shown as (l), and G-U base pairs as (:). (Reproduced, with permssion, from Maslov et al., 1992.)

■
━━━━━━━━━━━━━━━

MINICIRCLE kDNA AS A TARGET FOR DETECTION OF THE PARASITE

Because *T. cruzi* parasites occur in only small numbers in the blood of infected patients, they are difficult to detect. In the early 1980s, before the discovery of RNA editing and gRNAs, we reasoned that the kDNA minicircle, whatever its function, represented an appropriate molecular target for detection of this parasite by the technique of **hybridization**, because a large number of copies were present and they appeared to contain at least some short conserved sequences that would permit them to base pair with other similar molecules. In this technique (discussed also in Chapters 2 and 5), a radioactive single-stranded DNA fragment having a nucleotide sequence **complementary** to (able to base pair with) a short sequence on the minicircle, is allowed to form base pairs (that is, to hybridize) with kDNA molecules attached to a filter; the resultant "duplex" (double-stranded) molecules are detectable by their radioactivity.

Amplification of DNA: The Polymerase Chain Reaction

A chronically ill Chagas'-diseased patient has too few parasites for direct detection of parasite minicircle DNA by hybridization. However, at that time, Mullis and collaborators (Saiki et al., 1988) had just developed the **polymerase chain reaction (PCR)** technique by which small fragments of DNA are amplified (repeatedly replicated into enormous numbers of copies) in a test tube. This can be done if some nucleotide sequence information is already known. This technique has revolutionized molecular biology, and Mullis was awarded the Nobel Prize in 1993 for its discovery. Mullis claims that he got the idea for this novel technique while he was driving through the mountains of California with a female friend late one night, and he stopped the car to think it through (Mullis, 1990). Perhaps this proves the worth of leaving the lab once in a while and taking a drive with a friend!

The PCR technique depends on the ability to synthesize short DNA fragments of known nucleotide sequence and on the availability of a particular enzyme, **DNA polymerase**, which elongates a short DNA fragment in one direction by copying the nucleotide sequence from the other strand. In a short time, the method was improved by the use of a different DNA polymerase enzyme, one isolated from a high-temperature-adapted **thermophilic bacterium** that could survive repeated cycles of heating and cooling. The method is deceptively simple and consists of hybridizing two short synthetic single-stranded DNA molecules (called **primers**) onto the sequence to be amplified; adding new nucleotides to the 3' end of each molecule by use of the heat-stable DNA polymerase; separating the strands by heating; and repeating this replication process many times (Figure 6.13). After 20 such cycles, the small DNA fragments terminated by the original primers will be amplified 2^{20}, that is, approximately one million times. As few as 10 molecules, even a single molecule, can be amplified and then detected by any of several hybridization-based methods, using either radioactivity or a nonradioisotopic visualization procedure. The main problem with this method is that it is too sensitive—any slight contamination of the reagents used can easily lead to amplification of the contaminating DNA rather than the DNA actually sought. Another problem is the occurrence of mutations, changes in the DNA nucleotides that may occur during the amplification process itself.

However, in spite of these problems, this method has led to a revolution in many fields of modern research. In the 1960s, for example, the only way to amplify DNA sequences was to insert them into bacteria, grow the bacteria through many generations, then extract

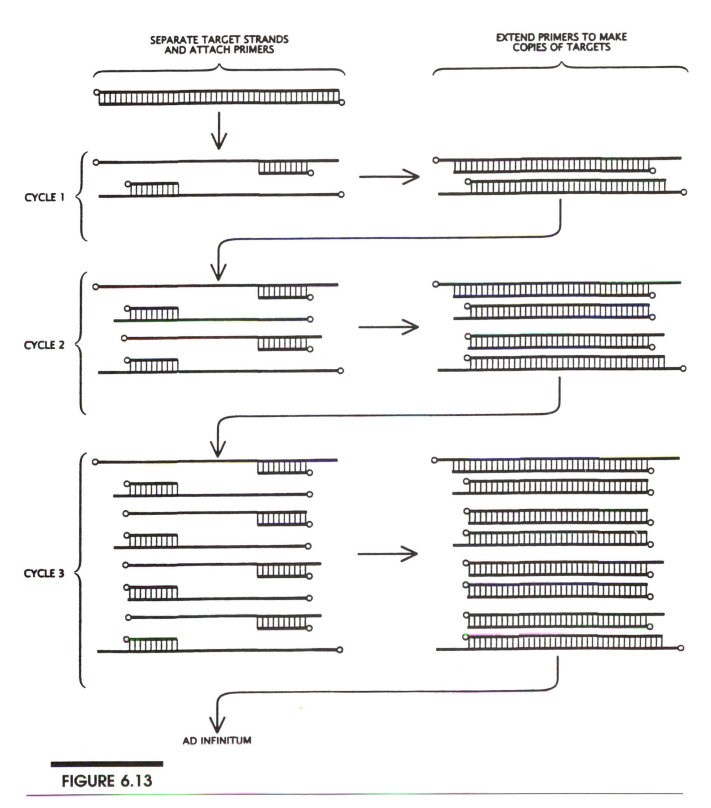

SEPARATE TARGET STRANDS AND ATTACH PRIMERS

EXTEND PRIMERS TO MAKE COPIES OF TARGETS

CYCLE 1

CYCLE 2

CYCLE 3

AD INFINITUM

FIGURE 6.13

Polymerase chain reaction (PCR). The first three cycles of a PCR treatment are diagrammed, out of 20 to 30 such cycles that are normally carried out. The number of DNA molecules in the system is doubled with each cycle. The first step (upper left) is to separate the strands of double-stranded DNA by heating, and to then attach oligonucleotide primers to them. A heat-stable DNA polymerase is then used to extend the primers by adding nucleotides to one of their ends. (Reproduced, with permission, from Kary B. Mullis, The Unusual Origin of the Polymerase Chain Reaction. Copyright © 1990 by Scientific American, Inc. All rights reserved.)

the DNA they had produced. Now, a DNA sequence can be amplified rapidly without using bacteria at all—just enzymes in a test tube! As trumpeted in newspapers and magazines, PCR has come to play a major role in forensic science, in which the presence of a particular individual at the scene of a crime can be demonstrated by the amplification of distinctive nucleotide sequences from the DNA that occurs in a single hair follicle or in a minute spot of blood or some other biological material (Blake et al., 1992). In paleontology (discussed also in Chapter 3), workers have been able to amplify DNA sequences from fossils millions of years old for direct comparison of their DNA with that of their living relatives.

Ancient DNA

The first ancient DNA was amplified in 1984 from a museum specimen of an animal called a quagga, a relatively recently extinct horse-like creature (Higuchi et al., 1984). A fragment of the mitochondrial genome of this specimen was PCR-amplified and sequenced. Soon thereafter, DNA fragments were amplified from 2,400-year-old Egyptian mummies (Pääbo, 1985) and 5,500-year-old human bone (Hanni et al., 1990). These discoveries were soon surpassed by amplification of DNA fragments from 18-million-year-old *Magnolia* leaves (Golenberg et al., 1990), 25- to 30-million-year-old fossil termites in amber (DeSalle et al., 1992), and recently, 120- to 135-million-year-old fossil weevils (Cano et al., 1993), also preserved in amber. There are, of course, cautionary notes being sounded, as discussed in Chapter 3, and it is possible that some of these reports may represent amplification of contaminating DNA from modern organisms, a well-known problem of PCR (Pääbo, 1989). It has also been argued that it is theoretically implausible for the chemical structure of DNA to survive for millions of years without being completely degraded (Lindahl, 1993). The evidence, however, seems strong that a number of these sequences actually have been obtained from fossilized DNA, because phylogenetic reconstructions derived from these data agree well with the known paleontological identity of the fossils.

PCR Amplification of *Trypanosoma cruzi* kDNA: The Chilean Mummy Connection

My interest in applying PCR to the detection of *T. cruzi* in biological materials was initially stimulated by a phone call from Christian Orrega, who was working in the laboratory of Alan Wilson, a pioneer in the amplification of ancient DNA (Pääbo et al., 1988). Orrega had access to tissues from 2,500-year-old desiccated mummies from Chile. The tissues reportedly showed at autopsy examination (now long delayed!) observable manifestations of Chagas' disease (Rothhammer et al., 1985). Orrega asked whether I would develop a PCR-based method to detect *T. cruzi* in such ancient tissue if he supplied me with tissue samples. Intrigued, I agreed to the project.

To use the PCR technique for amplifying parasite DNA, we first had to find appropriately conserved (that is, evolutionarily little changed) sequences with which the short PCR **oligonucleotide** primers could hybridize. Minicircle molecules were cloned and sequenced from three different strains of *T. cruzi*, and we found that the 1,500 base pair (bp) molecules were organized into four conserved and four variable regions, as shown in Figure 6.14 (Degrave et al., 1988). The conserved regions in other kinetoplastid protozoal species were shown to be sequences involved in DNA replication, and the variable regions, at least in minicircle molecules from the lizard parasite, *L. tarentolae*, and the African pathogenic trypanosome, *T. brucei*, to encode gRNA genes (Sturm and Simpson, 1991; Sturm and Simpson, 1990). When we aligned the sequences of 20 regions from five

A

```
                 10        20        30  │   40    S67 →  50        60        70      S33A →   90       100          110       120       130    ← S34A      150       160
     Consensus AAAattGGGq NtNNGAAATT cNGGAAANTN TGGTTTTGGG AGGGGCGTTC AAaTTTtGGG gCGgAAATTC ATGCATCTCC CCCGTACATT ATTTtGgCNA AAATGgGGAT TTTTcaNGGG AGGT-GGGGTT CGATTGGGGT TGGTGTAATA TAGNNANtNN NNTGg
```

cl1 cst 1 `....GGA..T A.AAA.TT.C TG..G..T.C .......... .......... ..C..... ...CAG.... .......... ........... ........GT. .......... ...-..C... ....G.... .......... .......... ...TATACTT GCGTA`
cl1 cst 2 `.....A.... -.TA..... .G....TG.T G.T..... .......... ..C..... .......... .......... ..........C. ....CCTA.. ...-G.... ....-..... .......... .......... ...TA.T.GA TTCAT`
cl1 cst 3 `...T..... -.TA..... .G....G.- ..A..... .A..... ..C..... .......... .......... ..........C. ....CCTA.. ...-ACG... ....-..... .......... .......... ...-CA.C.GG TA...`
cl1 cst 4 `....T....TT ACT-....A. .C....T.C .......... .......... ..C..... .......... .......... ..........CGG. TTT...AT. ...-ACG... ....-..... .......... .......... ...ACGGGTT GGATT`

cl2 cst 1 `.T..AG.... T.CG..... TC....A.T .......... .......... ..C..... ...CAG.... .......... ..........C...C. ....ACTA.. ...-G.... .......... .......... .......... ...CATGGGT TGC..`
cl2 cst 2 `..TT....TA CCTA...T.. TG..G..T.T GT..... .......... ..C..... .......... .......... ..........AC.C. ....AC.... ...-..C... .......... .......... .......... ...GC.TATA GA.TT`
cl2 cst 3 `..TT...T.. TCAA....A .G.AG..A-C .......... .......... ..C..... .......... .......... ..........C. ....CTA.. ...T-G... ....-..... .......... .......... ...GATC.GA TT...`
cl2 cst 4 `.........T A.CG....A. TC..G..A.C .T.A..... .......... ..C..... .......... .......... ..........C.C. ....T..... ...-..C... ....G.... .......... .......... ...GC.T.GG TC..A`

y01 cst 1 `....-G.... T.TG.G.... .-..GGTGGA .......... .......... ...G..C.C..A..A.. .......... ..........G.T.G. .......-. .G..T.C... ....-..... .......... .......... ...TAG.C.AG AT...`
y01 cst 2 `..-CGGA..T C.GTA.TT.. GT.A...C.G .......... .......... ...G..C CG.A..A.. .......... ..........G.T.G. .......G. .G..T.C... ....G.... .......... .......... ...-AG-CAAG AG...`
y01 cst 3 `...TTAG...T TCCGA....A GGAA..TCCT .......... .......... ...G..C CG.A..A.. .......... ..........G.T.G. .......-. .G..T.C... .......... .......... .......... ...GG.T.AT GG...`
y01 cst 4 `.C.T.AT.A. GGTG..... TC.A...TGT .......... .......... ...G..C CG.A..A.. .......... ..........G.C.G. .......G. .G..T.C... .......... .......... .......... ...GC.C.AT GTGT.`

y02 cst 1 `T.TGGATCCA C.GG....A. .A...GG.G .......... .......... ...G..C C..A..A.. .......... ..........GAT TTTA...G. .G..T.T... ....-..... .......... .......... ...-CT.C.-G AA.A.`
y02 cst 2 `....G...TA CCCT..... .G.T...T.A .A..... .......... ...G..C CG.A..A.. .......... ..........GAT TTTA...G. .G...AT. ....G.... .......... .......... ...-AGTTAGG TA...`
y02 cst 3 `TT.T....A T.AG..... .CT...TCT GT..... .......... ..C..... ...C..... .......... ..........GT..C. ....CCC..A .......-... ....-..... .......... .......... ...AGGG.TT TGA..`
y02 cst 4 `GT...AA... G.TGAG... TG.TT..G.C GT..... .......... ..C..... ...C..... .......... ..........CC..C. ....CTA..A .......-... ....G.... .......... .......... ...-AGGC.GG TGACT`

amp cst 1 `.G..A.T.TC CGGAA..... .CA...A.C CT.A..... ......GCG. .G...... ...CG..... .......... ..........A ........TAT TTTA...G. .G.CACGCC. ....-..... .......... .......... ...AC.GATT GT.CA`
amp cst 2 `G.T...A.. GCTAA..T.A .GT...TG.T GT..... ......GCG. ......CA. ..CT..... .......... ..........A ........ATCGA. TTT-.T... ....ACGA. ....-..... .......... .......... ...AC.GAGT GTG..`
amp cst 3 `..T....A.. T.AAA..T.C AC....G.T G.T..... ......GCG. ......G... C...... .......... ..........A ........GAT TTTG.C..A ....-G... ....-..... .......... .......... ...GCGTGGG TG..A`
amp cst 4 `.TGTAACAT. TCTG....A. TCT.G..AGC .T..... ......GCG. ......G... T...A.. .......... ..........CC.GAT TTTG.C..A ....-G... ....-..... .......... .......... ...TG.T.TG AT..C`
```

with **← S35** above (near position 80–90 region) and **S36 →** above (near position 140).

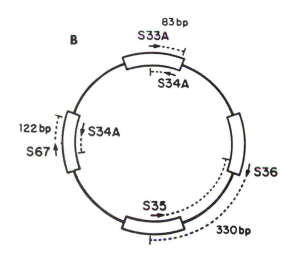

**B**

83 bp
S33A
S34A
122 bp
S67
S34A
S36
S35
330 bp

---

■■■■■■■■
## FIGURE 6.14

Alignment of conserved regions in five minicircles from three strains of *Trypanosoma cruzi*. Part A: Alignment of sequences showing highly conserved and less conserved regions. The locations of the primers (for example, "S34A") used for PCR are indicated by underlining, and their polarity by arrows. Dots indicate that identical nucleotides occur in all sequences; gaps are indicated by (-). The consensus sequence (listed at the top of the figure) shows well-conserved nucleotides as A, C, G, T, and less-conserved nucleotides as M, N, and a, c, g, and t. Part B: The circular diagram shows a minicircle having four conserved regions (indicated by boxes) and the localization of primers used for PCR. To simplify the diagram, only a single PCR product is shown for each conserved region.

sets of minicircles, we identified several highly conserved sequences to serve as primers for PCR amplification, either of 80 bp or 120 bp fragments within the conserved region, or of 330 bp fragments containing the adjacent variable regions (Figure 6.14). As summarized in Figure 6.15, we showed that this method worked and was specific for as few as 10 to 100 minicircle molecules from *T. cruzi* (Sturm et al., 1989), and that the method
```

FIGURE 6.15

Sensitivity of the PCR amplification of *Trypanosoma cruzi* minicircle DNA. kDNA was digested in guanidine blood lysate with a chemical nuclease (orthophen-anthraline-Cu++) and diluted with kDNA-free lysate to obtain known numbers of molecules as a substrate for each amplification. The upper panel shows that after staining the DNA with a fluorescent dye (ethidium bromide), a minimum of 10,000 minicircles can be detected. With a more sensitive technique for visualizing DNA (the Southern blot method), as few as 30 minicircles can be detected. The lower panel shows an autoradiograph of a Southern blot of this gel hybridized with a radioactive oligonucleotide probe for the minicircle conserved region. The identity of the rapidly migrating band seen in the lower panel in the vertically oriented lanes labeled 10,000, 3,000, and 1,000 is not known. Lanes C1 and C2 (at right) show results obtained for negative control amplifications, carried out without the DNA substrate, to check for contamination of reagents. (Reproduced, with permission, from Avila et al., 1991.)

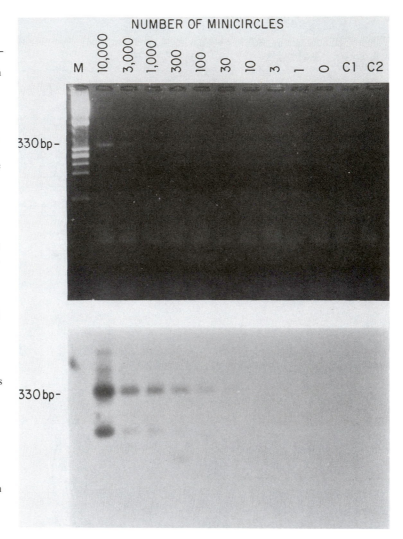

was unaffected by a several billionfold excess of human DNA (Figure 6.16). We then showed that amplification of these fragments could be accomplished for minicircles extracted from numerous strains of *T. cruzi* isolated from diverse regions throughout Latin America (Avila et al., 1991).

Diagnosis of Chagas' Disease in Patients by PCR Amplification of Minicircle kDNA

At this point, this method seemed more appropriate to detect *T. cruzi* parasites in live chronically ill patients, rather than in Chilean mummy tissues. It would be important to know whether live parasites of known strains were circulating in chronic Chagasic patients, or whether the disease was some sort of autoimmune phenomenon that persisted long after the parasites had disappeared.

To apply the PCR method to detect parasite DNA in patients, we needed a method to isolate parasite DNA from human blood. This presented a problem because blood contains a substance that inhibits PCR; we would have to use relatively large quantities (5 to 10 ml) of blood to detect the low numbers of parasites expected to be circulating in the veins of chronically ill patients. Herbert Avila, a graduate student in my laboratory, found that mixing blood with an equal volume of a particular concentrated salt solution (5 M guanidine-0.2 M ethylenediaminetetraacetic acid (EDTA); Avila et al., 1992) would **lyse** (disrupt and solubilize) the parasites as well as the host blood cells, releasing their DNA molecules into the medium, and would preserve the DNA against degradation even at or above room temperature for several weeks (Figure 6.17). This was an important break-through because samples could thus be taken in the field and transported to the labora-tory without refrigeration.

Because of the unique molecular organization of the parasite kinetoplast DNA, a method had to be found to release the catenated (interlocked) minicircles from the network so that the sensitivity of the procedure could be increased. Otherwise, a single network of DNA from a single trypanosome in even a 10 to 20 ml blood sample would inevitably be lost

FIGURE 6.16

Lack of effect of several billion-fold excess of human DNA on amplification of the 122 bp *Trypanosoma cruzi* minicircle fragment. The number of PCR cycles is indicated above each vertically oriented lane. The upper panel shows the stained gel, and the lower panel the autoradiograph of a Southern blot of this gel hybridized with a radioactive probe for the minicircle conserved region (see Figure 6.5). Primer controls (upper right) are control samples to which no DNA has been added. (Reproduced, with permission, from Sturm et al., 1989.)

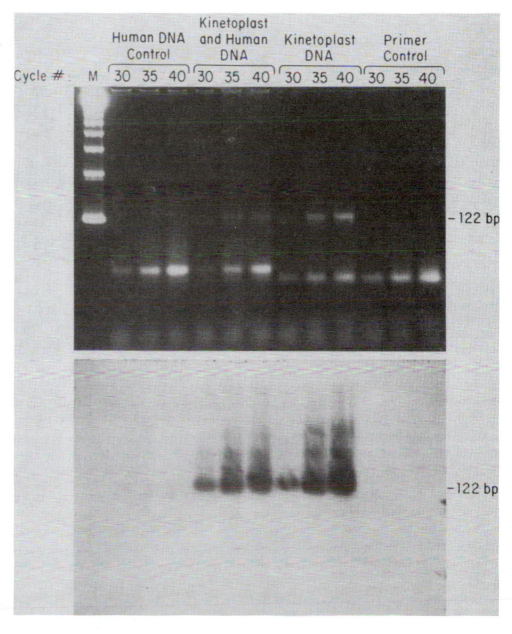

Thermal stability of closed circular DNA in the guanidine blood lysate at two temperatures. The positions of three types of "Bluescript plasmid" DNA molecules (closed circular, nicked circular, and linear) are shown. Note that closed circular DNA is stable for several weeks at 37°C. (Reproduced, with permission, from Avila et al., 1991.)

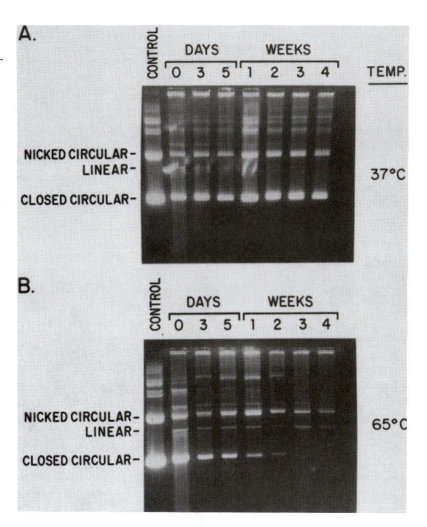

during the DNA isolation process. To solve this problem, we used a chemical **nuclease** (in particular, orthophenanthraline-Cu++) to cut, on average, each minicircle molecule once and thereby release 10,000 linear minicircle-derived molecules into the lysate (Avila et al., 1992). Recently, our Brazilian collaborators simplified and improved this procedure by simply boiling the lysate to break the minicircles, rather than by using the chemical nuclease (Britto et al., 1993). Release of the minicircles from the network allows us to work with a very small aliquot of the lysate (equivalent to only a few drops of blood) that contain a minimum of several hundred DNA molecules even if only a single trypanosome was originally present. Working with such a small sample facilitates isolating the total DNA from the blood and subsequently performing PCR amplification of parasite minicircle DNA fragments. The human host DNA (derived from white blood cells) is also isolated together with parasite DNA, and the mix can therefore be used for PCR amplification of any human genetic loci of interest, such as genes that cause genetic diseases or genes that confer resistance to various other diseases.

This method has been tested on blood samples from 95 patients from Brazil, many of whom had symptoms of chronic Chagas' disease and who had been previously tested by both xenodiagnosis and serological methods for the presence of the infection (Avila et al., 1993). We showed that the PCR method was approximately twice as sensitive as xenodiagnosis, and that it gave results that agreed well with serological assays that measured the presence of antibodies against the parasite (Table 6.1).

Amplification of the 330 bp variable region fragment of minicircle DNA yields one additional piece of information. Because these fragments are derived from all the minicircles in the network, and because the nucleotide sequences in these minicircle variable regions differ between different parasite schizodemes, the amplified DNA permits the specific strain of parasite present in a patient to be classified into an identified group. This may prove important, because different strains of the parasites may cause different clinical disease syndromes and may be susceptible to treatment with differing drugs. Such groupings should also allow us to learn which strains of the parasite are present in patients, insects, and other animals from differing geographical regions.

■

A MODEST PROPOSAL

Our scientific odyssey began with Charles Darwin's mysterious disease and continued with our recent research on *T. cruzi* and PCR, which actually was initially stimulated by a request by a colleague to examine the possibility of the presence of these parasites in 4000-year-old Chilean mummies, but which was sidetracked by studies of the presence of parasites in chronically ill Chagasic patients. In spite of these meanderings, it has not escaped our attention that the PCR amplification method we have developed to detect small numbers of parasites in chronic Chagasic patients might in fact be used to settle the question of Darwin's illness. I would like to end this chapter with a "modest proposal," namely, that we remove a sample of Charles Darwin for PCR analysis of *T. cruzi*. After all, he was interred in a very accessible location—in Westminster Abbey, right next to Isaac Newton—and, we only need a little piece!

TABLE 6.1

Comparison of results for blood samples analyzed by serologic, xenodiagnostic, and PCR tests for Chagas' disease.*

Sample Type	Number of Samples	Number of PCR Results	
		Positive	Negative
Clinically diagnosed Chagas' disease and			
Serology positive	91	91	0
Serology negative	1	1	0
Serology positive and			
Xenodiagnosis positive	48	48	0
Xenodiagnosis negative	35	35	0
Serology-negative nonchagasic patients			
from Virgin de Lapa, Brazil	3	2	1
UCLA Blood Bank donors	18	0	18

*The Chagasic samples are from patients from Virgin de Lapa, Minas Gerais, Brazil. The serological tests represent several standard immunological assays for the presence of antibodies to *T. cruzi*. Note that 35 patients who were xenodiagnosis-negative but serologically positive also proved positive by PCR, indicating that PCR was approximately twice as effective in detecting parasites as xenodiagnosis. In addition, two of the three patients who were serologically negative proved positive by PCR, suggesting either that PCR was more efficient than serology or that these were false-positive PCRs. All 18 University of California, Los Angeles (UCLA) blood bank donors proved negative by PCR.

Source: Reproduced, with permission, from Avila et al., 1993.

■

REFERENCES

Adler, D. 1989. Darwin's illness. *Isr. J. Med. Sci. 25:* 218–221.

Adler, S. 1959. Darwin's illness. *Nature 184:* 1102–1103.

Avila, H., Goncalves, A.M., Nehme, N.S., Morel, C.M., and Simpson, L. 1991. Schizodeme analysis of *Trypanosoma cruzi* stocks from South and Central America by analysis of PCR-amplified minicircle variable region sequences. *Mol. Biochem. Parasitol. 42:* 175–188.

Avila, H.A., Sigman, D.S., Cohen, L.M., Millikan, R.C., and Simpson, L. 1992. Polymerase chain reaction amplification of *Trypanosoma cruzi* kinetoplast minicircle DNA isolated from whole blood lysates: Diagnosis of chronic Chagas' disease. *Mol. Biochem. Parasitol. 48:* 211–222.

Avila, H.A., Pereira, J.B., Thiemann, O., De Paiva, E., Degrave, W., Morel, C.M., and Simpson, L. 1993. Detection of *Trypanosoma cruzi* in blood specimens of chronic chagasic patients by polymerase chain reaction amplification of kinetoplast minicircle DNA: Comparison with serology and xenodiagnosis. *J. Clin. Microbiol. 31:* 2421–2426.

Benne, R., Van den Burg, J., Brakenhoff, J., Sloof, P., Van Boom, J., and Tromp, M. 1986. Major transcript of the frameshifted coxII gene from trypanosome mitochondria contains four nucleotides that are not encoded in the DNA. *Cell 46:* 819–826.

Bernstein, R.E. 1984. Darwin's illness: Chagas' disease resurgens. *J. Roy. Soc. Med. 77:* 608–609.

Blake, E., Mihalovich, J., Higuchi, R., Walsh, P.S., and Erlich, H. 1992. Polymerase chain reaction (PCR) amplification and human leukocyte antigen (HLA)-DQ alpha oligonucleotide typing on biological evidence samples: Casework experience. *J. Forensic Sci. 37:* 700–726.

Blum, B., Bakalara, N., and Simpson, L. 1990. A model for RNA editing in kinetoplastid mitochondria: "Guide" RNA molecules transcribed from maxicircle DNA provide the edited information. *Cell 60:* 189–198.

Blum, B., and Simpson, L. 1992. Formation of gRNA/mRNA chimeric molecules *in vitro*, the initial step of RNA editing, is dependent on an anchor sequence. *Proc. Natl. Acad. Sci. USA 89:* 11944–11948.

Britto, C., Cardoso, M.A., Wincker, P., and Morel, C.M. 1993. A simple procedure for the physical cleavage of *Trypanosoma cruzi* kinetoplast DNA present in blood samples and its use on PCR-based diagnosis of chronic Chagas disease. *Mem. Inst. Oswaldo Cruz 88:* 171–172.

Cano, R.J., Poinar, H.N., Pieniazek, N.J., Acra, A., and Poinar, G. O., Jr. 1993. Amplification and sequencing of DNA from a 120- to 135-million-year-old weevil. *Nature 363:* 536–538.

Cech, T. R. 1991. RNA editing: World's smallest introns. *Cell 64:* 667–669.

Corell, R.A., Feagin, J.E., Riley, G.R., Strickland, T., Guderian, J.A., Myler, P.J., and Stuart, K. 1993. *Trypanosoma brucei* minicircles encode multiple guide RNAs which can direct editing of extensively overlapping sequences. *Nucleic Acids Res. 21:* 4313–4320.

Degrave, W., Fragoso, S., Britto, C., Van Heuverswyn, H., Kidane, G., Cardoso, M., Mueller, R., Simpson, L., and Morel, C. 1988. Peculiar sequence organization of kinetoplast DNA minicircles form *Trypanosoma cruzi. Mol. Biochem. Parasitol. 27:* 63–70.

DeSalle, R., Gatesy, J., Wheeler, W., and Grimaldi, D. 1992. DNA sequences from a fossil termite in Oligo-Miocene amber and their phylogenetic implications. *Science 257:* 1933–1936.

Feagin, J.E., Shaw, J.M., Simpson, L., and Stuart, K. 1988. Creation of AUG initiation codons by addition of uridines within cytochrome b transcripts of kinetoplastids. *Proc. Natl. Acad. Sci. USA 85:* 539–543.

Goldstein, J.H. 1989. Darwin, Chagas' disease, mind, and body. *Perspec. Biol. Med. 32:* 586–601.

Golenberg, E. M., Giannasi, D.E., Clegg, M.T., Smiley, C.J., Durbin, M., Henderson, D., and Zurawski, G. 1990. Chloroplast DNA sequence from a miocene Magnolia species. *Nature 344:* 656–658.

Hanni, C., Laudet, V., Sakka, M., Begue, A., and Stehelin, D. 1990. Amplification of mitochondrial DNA fragments from ancient human teeth and bones. *C. R. Acad. Sci. (Paris) III. 310:* 365–370.

Higuchi, R., Bowman, B., Freiberger, M., Ryder, O.A., and Wilson, A. C. 1984. DNA sequences from the quagga, an extinct member of the horse family. *Nature 312:* 282–284.

Kirchoff, L.V., Gam, A.A., and Gillan, F.C. 1987. American trypanosomiasis (Chagas' disease) in Central American immigrants. *Am. J. Med. 82:* 915–920.

Kirchoff, L.V. 1989. Is *Trypanosoma cruzi* a new threat to our blood supply? *Ann. Intern. Med. 111:* 773–775.

Lindahl, T. 1993. Instability and decay of the primary structure of DNA. *Nature 362:* 709–715.

Maslov, D.A., and Simpson, L. 1992. The polarity of editing within a multiple gRNA-mediated domain is due to the formation of anchors for upstream gRNAs by downstream editing. *Cell 70:* 459–467.

Morel, C., Chiari, E., Camargo, E., Mattei, D., Romanha, A., and Simpson, L. 1980. Strains and clones of *Trypanosoma cruzi* can be characterized by restriction endonuclease fingerprinting of kinetoplast DNA minicircles. *Proc. Natl. Acad. Sci. USA 77:* 6810–6814.

Mullis, K.B. 1990. The unusual origin of the polymerase chain reaction. *Scientific American 262:* 56–65.

Pääbo, S. 1985. Molecular cloning of ancient Egyptian mummy DNA. *Nature 314:* 644–645.

Pääbo, S. 1989. Ancient DNA: Extraction, characterization, molecular cloning, and enzymatic amplification. *Proc. Natl. Acad. Sci. USA 86:* 1939–1943.

Pääbo, S., Gifford, J.A., and Wilson, A.C. 1988. Mitochondrial DNA sequences from a 7000-year old brain. *Nucleic Acids. Res. 16:* 9775–9787.

Pollard, V.W., Rohrer, S.P., Michelotti, E.F., Hancock, K., and Hajduk, S.L. 1990. Organization of minicircle genes for guide RNAs in *Trypanosoma brucei. Cell 63:* 783–790.

Prata, A. 1981. *Carlos Chagas—Coletanea de Trabalhos Cientificos.* Editora Universidade de Brasilia, Brasilia.

Rothhammer, F., Allison, M.J., Nunez, L., Standen, V., and Arriaza, B. 1985. Chagas' disease in pre-Columbian South America. *Amer. J. Phys. Anthropol. 68:* 495–498.

Saiki, R.K., Gelfand, D.H., Stoffel, S., Scharf, S.J., Higuchi, R., Horn, G.T., Mullis, K., and Erlich, H.A. 1988. Primer-directed enzymatic amplification of DNA with a thermostable DNA polymerase. *Science 239:* 487–491.

Schmidtmeyer, P., 1824. *Travels into Chile.* (London: Longman, Hurst, Rees, Orme, Brown and Green).

Schmunis, G.A. 1991. *Trypanosoma cruzi*, the etiologic agent of Chagas' disease: Status in the blood supply in endemic and nonendemic countries. *Transfusion 31:* 547–557.

Shaw, J., Feagin, J.E., Stuart, K., and Simpson, L. 1988. Editing of mitochondrial mRNAs by uridine addition and deletion generates conserved amino acid sequences and AUG initiation codons. *Cell 53:* 401–411.

Simpson, L. 1972. The kinetoplast of the hemoflagellates. *Int. Rev. Cytol. 32:* 139–207.

Simpson, L. 1987. The mitochondrial genome of kinetoplastid protozoa: Genomic organization, transcription, replication and evolution. *Ann. Rev. Microbiol. 41:* 363–382.

Simpson, L., Maslov, D.A., and Blum, B. 1993. RNA editing in *Leishmania* mitochondria. In: R. Benne (Ed.), *RNA Editing* (New York: Ellis Horwood), pp. 53–85.

Stuart, K. 1993. RNA editing—the alteration of protein coding sequences of RNA. In: R. Benne (Ed.), *RNA Editing* (New York: Ellis Horwood), pp. 25–52.

Sturm, N.R., Degrave, W., Morel, C., and Simpson, L. 1989. Sensitive detection and schizodeme classification of *Trypanosoma cruzi* cells by amplification of kinetoplast minicircle DNA sequences: Use in diagnosis of Chagas Disease. *Mol. Biochem. Parasitol. 33:* 205–214.

Sturm, N.R., and Simpson, L. 1990. Kinetoplast DNA minicircles encode guide RNAs for editing of cytochrome oxidase subunit III mRNA. *Cell 61:* 879–884.

Sturm, N.R., and Simpson, L. 1991. *Leishmania tarentolae* minicircles of different sequence classes encode single guide RNAs located in the variable region approximately 150 bp from the conserved region. *Nucleic Acids Res. 19:* 6277–6281.

Thielen, J. 1991. *Science Heading for the Backwoods, Images of the Expeditions Conducted by the Oswaldo Cruz Institute to the Brazilian Hinterland 1911/1913.* (Rio de Janeiro: Fundaçao Oswaldo Cruz).

Tibayrenc, M., Ward, P., Moya, A., and Ayala, F. 1986. Natural populations of *Trypanosoma cruzi,* the agent of Chagas' disease, have a complex multiclonal structure. *Proc. Natl. Acad. Sci. USA 83:* 115–119.

Tibayrenc, M., Kjellberg, F., and Ayala, F.J. 1990. A clonal theory of parasitic protozoa: The population structures of *Entamoeba, Giardia, Leishmania, Naegleria, Plasmodium, Trichomonas,* and *Trypanosoma* and their medical and taxonomical consequences. *Proc. Natl. Acad. Sci. USA 87:* 2414–2418.

GLOSSARY

The definitions below are based on the way the words have been used in this book—some terms have additional meanings that are not provided below.

Å (Angstrom). A unit of length, one ten billionth (10^{-10}) of a meter.

Abdominal B **gene.** A gene occurring in the Antennapedia complex of homeotic genes in the fruit fly, *Drosophila*.

Abiotic. Of nonbiological origin.

Acetyl-CoA (acetyl-coenzyme A). A nonprotein chemical compound involved in aerobic respiration.

Active site. The part of an enzyme that interacts with the substrate upon which the enzyme acts.

Adenine (A). One of the four nitrogenous bases in the nucleotides of DNA and RNA.

Aerobic heterotrophy. Oxygen-consuming respiration.

Aerobic photoautotrophy. Oxygen-producing photosynthesis.

Aerobic respiration. The oxygen-consuming, energy-yielding process carried out by almost all eukaryotes; "breathing."

Aerobic. With reference to the presence of molecular oxygen.

Alanine. One of the 20 amino acids commonly occurring in proteins of living systems.

Albumin. Any of a group of simple water-soluble proteins that are coagulated by heat and are found in blood plasma, egg white, etc.

Alcaptonuria. A genetic disease in which the urine of the individual affected turns black on standing.

Alga. Any of diverse types of eukaryotic photoautotrophic "lower plants"; a seaweed.

Aligned sequences. See "Sequence alignment."

Allele. One of several forms of the same gene. An allele differs from other alleles by changes in its DNA sequence.

Alpha (α) helix. A coiled, helical conformation of a polypeptide (a long chain of amino acids) with maximal intrachain hydrogen bonding. A very common structure in proteins.

Alpha (α) hemoglobin. Together with beta (β) hemoglobin, one of the principal components of the oxygen-carrying molecule of blood.

Alu sequence. One of hundreds of thousands of copies of related sequences, each approximately 300 base pairs long, found scattered throughout the genomes of the great apes, among others.

Amber. Fossil resin, or sap, derived from trees.

Amino acid. A small molecule containing an amino group ($-NH_2$) and a carboxylic acid group ($-COOH$) that can be linked together like a string of beads to form a protein. In living systems there are 20 amino acids, each with a different, distinct side chain.

Amino acid code. See "Genetic code."

Amino group. The chemical group, $-NH_2$.

Ammonia. The chemical compound NH_3. A common biological waste product.

Amphibian. Any of various vertebrate animals, such as salamanders and frogs, that in the adult form can live on land but that require an aqueous environment for reproduction.

Anaerobic. With reference to the absence or near-absence of molecular oxygen (O_2).

Anaerobic heterotrophy. Non-oxygen-consuming fermentative metabolism.

Anaerobic photoautotrophy. Non-oxygen-producing photosynthesis.

Angstrom (Å). A unit of length, one ten billionth (10^{-10}) of a meter.

Anoxic. With reference to the complete absence of molecular oxygen (O_2).

Anoxygenic. With reference to a process that does not produce molecular oxygen.

Anoxygenic photosynthesis. Non-oxygen-producing photoautotrophy, such as that carried out by photosynthetic bacteria.

Antennapedia (*Antp*) **complex.** A region on chromosome 3 of the fruit fly *Drosophila* that contains three homeotic genes that define the identities of the anterior segments of the fly.

Anteroposterior. The front-to-rear axis of metazoans.

Antibody. A protein produced by the immune system that recognizes a specific foreign antigen (often a toxic chemical), and triggers the immune response which then attempts to destroy the foreign body.

Antigen. Any molecule whose entry into an organism provokes the synthesis of an antibody.

Apex chert. A fossiliferous horizon of the Apex Basalt, a 3,465-million-year-old geological unit of Western Australia.

Archaebacteria. A taxonomic group composed of non-nucleated (prokaryotic) microorganisms; together with Eubacteria and Eukaryotes, Archaebacteria constitute one of the three main groups of the Tree of Life.

Archean Eon. The earliest of the three principal divisions (eons) of geologic time, extending from the formation of Earth (approximately 4,500 million years ago) to approximately 2,500 million years ago.

Archezoan. Any eukaryote that lacks mitochondria and plastids (such as chloroplasts).

Arginine. One of the 20 amino acids commonly occurring in proteins of living systems.

Arthropoda. Any member of the most species-rich phylum of animals, characterized, in part, by a well formed exoskeleton and jointed limbs. Includes, among others, insects, chelicerates (such as spiders) and crustaceans (such as shrimp and crabs).

Asexual. With reference to organisms that lack the capability to reproduce sexually (that is, they lack life cycles in which meiosis and syngamy occur).

Aspartic acid. One of the 20 amino acids commonly occurring in proteins of living systems.

ATP (adenosine triphosphate). A compound present in all cells that provides energy derived from food or sunlight for energy-expending biologic processes.

ATPase. An enzyme that breaks down ATP. A large number of different enzymes have ATPase activity in addition to other functions.

Autapomorphy. Unique, derived characteristic; a characteristic unique to just one member of a lineage, or to just one lineage among many lineages.

Autonomic. The involuntary nervous system of mammals.

Autotrophy. The metabolic process characteristic of plants and plant-like organisms in which carbon dioxide serves as the primary source of carbon.

Bacteria. A general term for all prokaryotes, that is, the Archaebacteria and Eubacteria (though some exclude the cyanobacteria).

Bacteriochlorophyll. Any of several structurally similar, light-absorbing pigments that play a central role in bacterial photosynthesis.

Bacteriophage. A virus that infects bacteria. Phage is often used as an abbreviation.

Bacteriorhodopsin. A light-absorbing pigment that is involved in light-driven production of ATP in Archaebacteria of the genus *Halobacterium*.

Barrier effect. The winnowing effect of natural selection that determines whether mutational changes in proteins are accepted or rejected and, thus, whether the changes are carried over into subsequent generations.

Base pair. Any of the bonding pairs of nucleotides specified for "base pairing."

Base pairing. The capability of certain nitrogenous bases in nucleic acids to pair by hydrogen bonding; specifically, between the two strands of double-stranded DNA, adenine-thymine (A-T) and guanine-cytosine (G-C) bonding; between single-stranded RNA and one strand of DNA, U-A bonding between uracil (of RNA) and adenine (of DNA), A-T bonding between adenine (of RNA) and thymine (of DNA), and G-C bonding between guanine and cytosine; and between two single-stranded RNAs, uracil-adenine (U-A) and guanine-cytosine (G-C) bonding.

Base ratio. In DNA, the ratio of the nitrogenous bases guanine + cytosine (G + C) to adenine + thymine (A + T); in RNA, the ratio of G + C to the nitrogenous bases adenine + uracil (A + U).

Beta (β) hemoglobin. Together with alpha (α) hemoglobin, one of the principal components of the oxygen-carrying molecule of blood.

Bilateria. Encompasses those animal phyla that are triploblastic (have ectoderm, mesoderm, and endoderm). Includes almost all animals, but excludes, among others, sponges and cnidarians.

Biological Species Concept. See "Species."

Biomarker. A contraction of the term "biological marker compound." Refers to chemical fossils, such as cholesteranes, of biological origin.

Biosphere. The sum total of all living systems.

Biosynthesis. The process of manufacture of organic compounds by living systems.

Biosynthetic pathway. Any of numerous enzyme-mediated multistep processes by which organic compounds are formed in living systems.

Biota. The sum of all organisms of a particular region or period.

***bithorax* complex.** A region of chromosome 3 of the fruit fly *Drosophila* that contains three homeotic genes that define the identities of the posterior segments of the fly.

***bithorax* mutant.** A mutation in one of the genes occurring in the *bithorax* complex of the fruit fly, *Drosophila*.

Black smoker. A type of hydrothermal vent in the deep sea that discharges a continuous black plume.

Bond energy. Chemical energy that holds together the atoms of a molecule.

Bootstrap replicate. Bootstrapping is a technique used to determine the relative strengths of support for different evolutionary trees within a set of sequences. Each bootstrap replicate in a bootstrap analysis is a new set of sequences of the same length as the original set, generated by resampling, with replacement, the sites of the original set.

Cambrian Explosion. The name given to the sudden appearance of the major phyla in the fossil record during the Cambrian Period.

Cambrian Period. The earliest period of the Phanerozoic Eon of Earth's history, extending from approximately 545 million years ago to approximately 505 million years ago.

Cap structure. A modified nucleotide at the 5′ end of eukaryotic mRNAs.

Carapace. The tough, commonly mineralized surface covering, characteristic of some animals such as crabs, lobsters, and turtles.

Carbohydrate. A carbon-, hydrogen-, and oxygen-containing organic polymeric compound composed of sugar monomeric subunits.

Carbonaceous meteorite. A relatively rare type of stony meteorite (as opposed to iron/nickel meteorites) that contains about 5% organic compounds including inorganically produced amino acids.

Carboxylic acid group. The chemical group -COOH.

Capillary. The smallest type of blood vessel in the body. The capillaries are where most of the oxygen exchange occurs between the blood and tissues of an animal.

Carotene. An orange-colored, light-absorbing organic pigment that occurs in many types of photosynthetic bacteria, algae, higher plant leaves, and is characteristic of carrot roots.

Catalysis. The process of facilitating, increasing the rate of, a chemical reaction.

Catenation. Circular molecules that are interlocked, like rings in a chain.

Cellulose. The polymeric organic compound composed of glucose monomeric subunits that is characteristic of the walls of plant cells.

Cenozoic Era. The latest of the three Eras that make up the Phanerozoic Eon of Earth history, extending from 65 million years ago to the present.

Central dogma. The concept that the flow of genetic information goes from DNA to RNA to protein, but that it cannot flow from protein to either RNA or DNA.

Cervical vertebra. Backbone part of the neck region.

Chagas' disease. Incurable chronic disease that affects the heart and causes fever prevalent in South America. Caused by a eukaryotic protozoal parasite, *Trypanosoma cruzi*.

Chemical nuclease. Chemical complex that can chemically cleave a DNA molecule at random sites for example, using orthophen-anthraline-Cu++.

Chemosynthetic autotrophy. The metabolic process of organisms in which the oxidation of hydrogen (H_2), sulfur (S^0), or sulfide (S^{2-}) is used as the principal source of energy.

Chert. A type of rock composed of microcrystalline quartz, SiO_2.

Chlorophyll. Any of several structurally similar, light-absorbing pigments that play a central role in the oxygenic photosynthesis of cyanobacteria and plants.

Chloroplast. An organelle occurring in plant cells in which the process of photosynthesis occurs.

Cholesteranes. Molecules derived from the degradation of cholesterol that may be identified in the fossil record.

Cholesterol. A steroid-type organic compound important in various physiological processes.

Chordates. A phylum of animals characterized by, among other features, the presence of a dorsal notochord at some stage of development and a dorsal hollow nerve cord. Includes, among others, tunicates and vertebrates.

Chromista. A group of predominantly photosynthetic eukaryotes. Those with chloroplasts have them within the rough endoplasmic reticulum, not in the cytosol as in plants. Includes, among others, diatoms, brown algae, and kelps.

Chromoplast. A type of plastid that accumulates carotenoid pigments, and in many species of plant is responsible for yellow-orange-red coloration of the petals.

Chromosome. Elongate structures, in eukaryotes occurring in the cell nucleus, that contain the hereditary molecule, DNA. Informally, chromosome is also used to refer to the DNA in an organelle, or any noneukaryote.

Ciliate. Any of diverse unicellular protozoans characterized by the occurrence of numerous hair-like cilia.

Citric acid cycle. The electron transport cycle of aerobic respiration.

Clade. A branch, or lineage of organisms on an evolutionary tree.

Cladistics. An approach to systematics that only admits groups based on the presence of explicitly stated, shared, derived characteristics (synapomorphies).

Clone (verb). The insertion of a piece of foreign DNA into a plasmid or phage. A bacterium is infected with the plasmid or phage, which is then cultured so that large quantities of the foreign DNA, or the protein it codes for, can be recovered.

Cnidaria. Phylum of animals that include groups such as jellyfish, sea anemones, and corals.

Coding. The sequence of nucleotides in a nucleic acid molecule encodes the message that results in production of a particular sequence of amino acids in a synthesized protein molecule.

Codon. A triplet of nitrogenous base-containing nucleotides of the genetic (amino acid) code.

Coenzyme. One of numerous nonprotein chemical compounds required for function of various proteinaceous enzymes.

Colinearity. With reference to the same relative spatial distribution of genes and the expression of their products along the anteroposterior body axis.

Collagen. The primary protein constituent of connective tissue and bone. This fibrous protein is the most abundant protein in mammals, constituting 25% of their total protein.

Complementary. With reference to any two nitrogenous bases that are capable of forming base pairs (in DNA: G-C, and A-T; in RNA: G-C, and A-U).

Conifer. Any of various types of cone-bearing evergreen trees.

connectin **gene.** A gene in mammalian cells that produces a molecule that causes cells to adhere to each other and is involved in establishment of nerve connections in muscle tissue.

Cretaceous Period. The latest of the three Periods of the Mesozoic Era of Earth's history, extending from approximately 145 million years ago to 65 million years ago.

Crown group. See "Total group."

Cryptodire. One of two major lineages of living turtle characterized, in part, by a neck retraction mechanism that draws the neck vertically into the shell.

Cryptogene. A maxicircle gene, the transcripts of which are edited to produce translatable messenger RNAs.

Cyanobacteria. Eubacteria capable of oxygen-producing photosynthesis.

Cyst. A thick-walled protective membrane enclosing a cell, larva, or organism.

Cysteine. One of the 20 amino acids commonly occurring in proteins of living systems.

Cytochrome. A type of protein occurring in the electron transport chain of aerobic respiration.

Cytosine (C). One of the four nitrogenous bases in the nucleotides of DNA and RNA.

Cytoskeleton. A network of fibers in the cytoplasm (the region between the cell membranes and nucleus) of the eukaryotic cell.

Cytosol. The watery intracellular cytoplasmic fluid.

Dark reactions. Chemical reactions of photosynthesis that do not require light energy.

Darwinian struggle. The competition involved in natural selection.

Deformed **gene.** A gene occurring in the Antennapedia complex of homeotic genes in the fruit fly, *Drosophila*.

Degradative pathway. A series of biochemical reactions resulting in breakdown of a molecule to produce smaller molecules and, commonly, cellular energy.

Deuterostomia. Animal phyla characterized, in part, by the way they develop: the blastopore is located posteriorly and may become the anus. Includes, among others, echinoderms and chordates.

Developmental gene. A gene involved in the development of an organism (that is, from egg to adult).

Diatom. Any of a large number of unicellular algae occurring in marine or fresh water, each having a cell wall made of two halves impregnated with silica.

Dinocyst. Cysts of dinoflagellates.

Dinoflagellate. A type of flagellated photosynthetic protist that lives in the plankton.

Dinosterane. Geologically modified compounds similar to cholesterol that are believed to originate in the cell membranes of dinoflagellates.

Diploid. A cell or organism possessing two sets of chromosomes; hence, every gene occurs at least twice in a diploid organism, such as humans (unless the gene occurs on an unduplicated sex chromosome).

DNA (deoxyribonucleic acid). The genetic information-containing molecule of cells, a double-stranded nucleic acid made up of nucleotides that contain deoxyribose sugar and the nitrogenous bases adenine [A], thymine [T], guanine [G], and cytosine [C]. Each strand of DNA has a polarity, running from 5′ to 3′. The 5′ and 3′ refer to particular carbons of the sugar that, along with phosphate (PO_4) molecules, forms part of the backbone of DNA. DNA replication and transcription always run from 5′ to 3′.

DNA-binding protein. Any of various proteins that bind to DNA. These proteins are major regulators of gene expression.

DNA polymerase. An enzyme that polymerizes (links together) nucleotides to synthesize a DNA strand by copying the nucleotide sequence of a complementary strand.

DNA replication. A process that occurs during cell division in which the information-containing DNA is duplicated.

DNA self-copying. See "DNA replication."

Domain. A protein domain corresponds to a functional subunit of a protein. There appears to be some correspondence between domains and the exons of the genes that code for the protein, although the concept of a domain is sufficiently fuzzy such that this correspondence is not firmly established.

Duplex anchor. A double-stranded region created by hybridization (base pairing) of complementary sequences of the mRNA and a specific guide RNA, thought to initiate the process of RNA editing.

E. coli. Abbreviation for the bacterium *Escherichia coli*.

Echinodermata. Phylum of animals characterized by a unique form of calcitic skeleton (stereom), a unique circulatory system (the water vascular system), and, in living representatives and most extinct species, fivefold symmetry. Includes, among others sea stars and sea urchins.

Ecologic generalist. An organism, such as some cyanobacteria, that are capable of living in ecologically diverse habitats.

Ecology. The science that deals with the interrelationships among organisms inhabiting a common environment and with the relationships between organisms and their environment.

Ecosystem. The complex of a biological community and its environment functioning as an interactive ecological unit.

EDTA. Ethylenediaminetetraacetic acid used to inactivate enzymes, such as those that damage or modify DNA. The chemical works by removing divalent ions (usually Mg^{2+}) required by the enzymes to function.

Electron acceptor. In a chemical reaction, a molecule that accepts one or more electrons contributed by another molecule.

Electron carrier. In a chain of chemical reactions, molecules that accept electrons from an electron donor and pass those electrons on to an electron acceptor.

Electron donor. In a chemical reaction, a molecule that contributes one or more electrons to another molecule.

Elongation factor protein. A protein that associates with ribosomes during the addition of each amino acid to forming proteins.

Embryology. The science that deals with the early stages of development of an organism.

Endoparasite. A parasite that lives within the body of its host.

Endosymbiosis. An intracellular symbiotic relationship.

Endosymbiotic hypothesis. The hypothesis that the organelles of eukaryotes (for example, mitochondria and chloroplasts) were derived from free-living bacteria that were engulfed by an ancestral eukaryote.

Enzyme. A protein capable of catalyzing a biochemical reaction.

Eocyte. Any of a group of thermophilic, sulfur-metabolizing, Archaebacteria thought to be more closely related to the eukaryotes than to any other bacterial group.

Erythrocyte. A red blood cell of vertebrates.

Escherichia coli. A eubacterium used extensively in molecular biology, including use for cloning.

Estrogen. A steroid hormone that stimulates production of female secondary sex characteristics.

Eubacteria. A taxonomic group composed of non-nucleated (prokaryotic) microorganisms; together with Archaebacteria and Eukaryotes, Eubacteria constitute one of the three main groups of the Tree of Life.

Eukaryotes. A taxonomic group that consists of organisms that are composed of one or more nucleus-containing cells; together with Archaebacteria and Eubacteria, Eukaryotes constitute one of the three main branches of the Tree of Life.

Evolutionary distance. The degree of evolutionary relatedness between two biologic lineages.

Exon. Any of many portions of a gene that encodes information that will be translated into protein, or functional RNA.

Fabaceae. The pea or legume family of plants; the second or third largest of 300 families of flowering plants, with about 13,000 species. Includes predominantly herbs but also shrubs and some trees.

Family-box amino acid. Amino acids coded for by four different codons in which the same two nitrogenous bases occur at the first two positions in the codons. Any of the four bases may occur at the third position. There are eight family-box amino acids.

Farnesyl. A 15-carbon isoprenoid hydrocarbon composed of three isoprene monomeric subunits.

Felid. Any animal belonging to the cat family, Felidae, including the lion and domestic cat.

Fermentation. Anaerobic metabolism.

First-base position. In a codon, the first of the three positions in the triplet.

Five prime (5′) end, of DNA. See "DNA."

Flagellum. A whip-like structure, having a characteristic internal morphology constructed from microtubules, that functions to propel a cell through an aqueous medium.

Flatworm. Any of a number of unsegmented worms belonging to the phylum Platyhelminthes.

Free-living. With reference to metabolic independence, said of an organism that is neither parasitic nor symbiotic.

Gas chromatography/mass spectrometry (GCMS). A technique that may be used to obtain and identify molecules of biological origin in sedimentary rocks.

Gastrulation. The process involved in formation of a gastrula, an early stage of embryonic development.

Gel electrophoresis. A technique by which macromolecules are separated in a gel matrix by their size, shape, and electrical charge (and therefore, if DNA or RNA, by their lengths).

Gene. A segment of DNA involved in the production of a protein or RNA molecule; it includes regions preceding and following the coding region, as well as intervening sequences (introns) between individual coding segments (exons).

Gene conversion. Gene conversion can change the sequence of a gene, or portion of a gene, to that of a closely related copy (whether it be an allele or paralog).

Gene duplication. The process (like DNA replication) by which duplicate copies of genes are biosynthesized.

Genealogy. The genetically determined familial relationships among a group of organisms.

Genetic code. The three-"letter" code, in which each "letter" may be any of the four nitrogenous base- (adenine-, uracil-, guanine-, or cytosine-) containing nucleotides of mRNA, which contains the information required for the addition of the correct amino acid to the end of a protein during its synthesis (or that terminates protein manufacture).

Genetic drift. The tendency, particularly in small populations, for gene frequencies to change by chance, regardless of whether the gene is advantageous.

Genetic locus. A site on a chromosome occupied by a specific gene; more loosely, the gene itself and all its allelic states.

Genetic recombination. The exchange of genes on a chromosome of an organism into a combination differing from that of either of its parents.

Genetics. The science that deals with heredity and variation.

Genome. All the DNA contained within one set of chromosomes (for haploid organisms this is all the DNA). One can also talk of the genome of an organelle.

Genotype. The genetic constitution of an organism, in contrast with the appearance (the phenotype) of the individual.

Geranyl. A 10-carbon isoprenoid hydrocarbon composed of two isoprene monomeric subunits.

Globin. A blood protein involved in carrying oxygen.

Glucose. Blood sugar, used by cells as their main source of energy. A six-carbon sugar, $C_6H_{12}O_6$.

Glutamate dehydrogenase. A key enzyme in energy production.

Glutamic acid. One of the 20 amino acids commonly occurring in proteins of living systems.

Glycine. One of the 20 amino acids commonly occurring in proteins of living systems.

Glycolysis. A type of fermentation (anaerobic metabolism) in which each molecule of glucose is broken down to produce molecules of pyruvate and energy.

***goosecoid* gene.** A homeobox gene occurring in frogs which directs cell migration during an early stage of embryonic development.

Gram-positive bacteria. A taxonomic group of Eubacteria (including, for example, *Heliobacterium*) that is characterized by cell walls that become pigmented after treatment with a stain known as "Gram's stain."

Green gliding bacteria. A taxonomic group of Eubacteria that includes, for example, photoautotrophs of the genus *Chloroflexus*.

Green sulfur bacteria. A taxonomic group of Eubacteria that includes, for example, photoautotrophs of the genus *Chlorobium*.

gRNAs. See "Guide RNAs."

Guanidine. A salt used to dissociate proteins from nucleic acids in the purification of nucleic acids.

Guanine (G). One of the four nitrogenous bases in the nucleotides of DNA and RNA.

Guide RNAs (gRNAs). A small RNA molecule that encodes the sequence information for editing; gRNAs contain a sequence at their 5′ end that can base pair with the mRNA just adjacent to the region to be edited.

Halobacteria. A group of Archaebacteria that can live in high concentrations of salt.

Halophile. An organism exceptionally tolerant of high environmental salinity.

Haltere organ. The balancing organ apparatus of advanced flies, such as *Drosophila* (these are highly modified wings).

Haploid. A cell or organism possessing one set of chromosomes.

Hemoglobin. A protein of the red blood cell functioning in oxygen (O_2) transport.

Hemoglobin S allele. A defective allele of hemoglobin, that when homozygous, results in sickle cell anemia.

Heredity. Transmission of genetic factors that determine individual characteristics from one generation to the next.

Heterocyst. A specialized thick-walled type of cell characteristic of advanced cyanobacteria that encloses and protects the oxygen-sensitive nitrogenase enzyme system.

Heterogeneous nuclear RNA (hnRNA). RNA, having no known function, that occurs in the cell nucleus.

Heterotrophy. The metabolic process characteristic of animals and animal-like organisms in which organic compounds serve as the principal source of carbon, and thereby energy.

Heterozygous. Having a different allele of a gene (at a single locus) on each chromosome.

Histone. Any of a particular group of usually highly conserved proteins that are associated with the packaging of DNA into the chromosomes of eukaryotic cells.

Homeobox. The conserved part of some chromosomal genes that contains the code for the homeodomain.

Homeodomain. The evolutionarily conserved DNA-binding portion of the protein coded for by the homeobox.

Homeotic gene. A gene that, when disrupted, produces a homeotic transformation.

Homeotic transformation. Malformations in developing or regenerating animals in which a segment or region of the body is transformed into the likeness of some other normal body segment or region.

Homologous genes. Genes that share a common evolutionary (gene) ancestor.

Homologous proteins. Proteins produced by homologous genes.

Homologues. Structures (for example, bones) or molecules (for example, proteins) that are evolutionarily derived from a common ancestor.

Homoplasy. Similarities between species that are not a result of shared ancestry.

Homozygous. Having the same allele of a gene (at a single locus) on each chromosome.

Horizontal gene transfer. The transmission of genetic material between organisms (usually different species), other than by vertical descent.

Hox **code.** The biochemical information that instructs *Hox* protein-containing cells to assume appropriate locations in the developing vertebrate body.

Hox **genes.** Genes in vertebrate animals that are related (homologous) to the homeotic genes in fruit flies, that is, those belonging to the *Antennapedia* and *Bithorax* complexes.

Hybrid DNA. Double-stranded DNA in which each strand is derived from a different biological source (for example, a combination of two single strands, one from a rat and one from a mouse).

Hybridization. The process by which two complementary nucleic acid sequences can form a double-stranded molecule by the formation of base pairs.

Hydra. Any of numerous small tubular freshwater jellyfish-related hydrozoan polyps.

Hydrogen bond. A weak electrostatic attraction between one electronegative atom and a hydrogen covalently linked to a second electronegative atom. Hydrogen bonds hold the two strands of DNA together in double-stranded DNA, as well as α helices in proteins.

Hydrogen source. The chemical source of hydrogen (electrons) that during photosynthesis reacts with carbon dioxide to produce organic compounds.

Hydrophobic. Lack of affinity for water.

Hydroxyapatite. Mineral phase that provides bone with its hardness.

Hypobradytely. The range (distribution) of exceptionally slow rates of morphological evolutionary change characteristic of cyanobacteria.

Immunology. The study of the immune system (the system in vertebrates responsible for protecting the body against diseases by producing antibodies).

Informative site. A position in a gene where the pattern of nucleotides in a sequence alignment indicates that some species are more closely related to each other than to other species; a site that is a synapomorphy.

Infrared light. That portion of the electromagnetic spectrum lying outside the visible spectrum at its red end.

Insect vector. A host for a parasite during one stage of its life cycle, commonly a blood-sucking insect that transmits the parasite to a vertebrate host.

Insulin. A protein pancreatic hormone in humans.

Intron. Any portion of a gene that lies between two exons, i.e., that does not encode information that will be translated into protein, or functional RNA. Introns are transcribed into mRNA, but are removed before the mRNA is translated into protein.

Invertebrates. An informal taxonomic grouping that includes all animals that are not vertebrates.

Ion. An electrically charged atom or group of atoms.

Isoleucine. One of the 20 amino acids commonly occurring in proteins of living systems.

Isoprene. The five-carbon (C_5H_8) monomer that, when linked in linear sequence, makes up the isoprene rubber polymer.

Isotope. One of two or more atoms belonging to the same element and that have different masses because of a differing number of neutrons. Isotopes may be radioactive, or stable (nonradioactive).

Isotopic date. Age of a rock (or organic material if it is less than 70,000 years old) obtained by measuring the ratio of a radioactive isotope and one of its stable decay products.

Junk DNA. DNA that is not known to code for a protein or RNA product.

Jurassic Period. The second of the three Periods of the Mesozoic Era of Earth's history, extending from approximately 210 million years ago to approximately 145 million years ago.

Kb. See "Kilobase."

Keratin. A fibrous protein that occurs in the outer layer of the skin and in hair, nails, feathers, scales, hooves, etc.

Kilobase. A unit of length (1000 bases) used for DNA and RNA sequences.

Kilocalorie. A unit of energy, the amount of heat required to raise the temperature of 1 kilogram of water 1°C.

Kinetoplast DNA (kDNA). The network of catenated minicircles and maxicircles within the single mitochondrion of kinetoplastid protozoa.

Kinetoplastid protozoa. A large group of mainly parasitic protozoan species characterized by the presence of a large mass of mitochondrial kinetoplast DNA at the base of the flagellum.

labial **gene.** A gene occurring in the homeotic complex of the fruit fly, *Drosophila*.

Lactic acid. An organic acid, $C_3H_6O_3$, usually produced by fermentation.

Lateral gene transfer. See "Horizontal gene transfer."

Light-harvesting pigment. Organic compounds, such as chlorophyll, that absorb light energy in photosynthesis.

Light reactions. Chemical reactions of photosynthesis that require light energy.

Lobe-finned fish. Fish that have both bone and muscle in their limbs, in contrast to having simple fins as seen in most fish, such as perch or carp. There are only seven living species, the coelacanth and six species of lungfish. Amphibians and land vertebrates were derived from lobe-finned fish.

Locus. See "Genetic locus."

Long interspersed element (LINE). Long noncoding stretches of DNA found repeatedly in chromosomes.

Lumbar vertebra. Backbone part of the lower back region.

Lyse. The complete disruption and solubilization of cells.

Lysine. One of the 20 amino acids commonly occurring in proteins of living systems.

Macromolecule. Any large molecule, such as a protein, DNA, or RNA.

Maize. Another name for corn.

Malaria. A potentially fatal infectious disease characterized by recurring attacks of fever and chills, caused by the bite of an anopheles mosquito infected with protozoans of the genus *Plasmodium.*

Maxicircles. Large, circular, double-stranded DNA molecules that together with minicircles are linked into the single network of kinetoplast DNA and that encode ribosomal RNAs and structural genes for mitochondrial proteins.

Meiosis. Process of nuclear division that reduces the number of chromosomes from $2n$ to $1n$ in each of four product cells. Products may mature to germ cells (sperm or eggs).

Melting point of DNA. The temperature at which double-stranded DNA separates into its component strands.

Meristic series. A series of body segments.

Mesozoic Era. The second of three Eras that make up the Phanerozoic Eon of Earth's history, extending from approximately 245 million years ago to 65 million years ago.

Messenger RNA. See "mRNA."

Metabolic pathway. Any of several enzyme-mediated multistep processes by which metabolism is carried out in living systems.

Metabolism. The sum of the energy-requiring and energy-producing processes that occur during the chemical buildup and breakdown of organic compounds in living systems.

Metamorphism. A change in the constitution of a rock produced by pressure and heat.

Metazoan. A taxonomic category whose membership is restricted to multicellular animals.

Meteor. Solar system matter observable when it falls into Earth's atmosphere and is heated by friction to temporary incandescence. A "shooting star."

Meteorite. A meteor that reaches Earth's surface.

Methane. The colorless gaseous hydrocarbon, CH_4.

Methane-producing Archaebacteria. Archaebacterial prokaryotes that produce methane as a product of their metabolism.

Methanogen. Any of a group of Archaebacteria that produce methane.

Methionine. One of the 20 amino acids commonly occurring in proteins of living systems.

Microbe. Informal term for any of diverse types of eubacterial or archaebacterial prokaryotes.

Micrometer (μm). A unit of length, one millionth (10^{-6}) of a meter.

Microsatellite DNA. See "Satellite DNA."

Minicircles. Small, circular, double-stranded DNA molecules that together with maxicircles are linked by catenation into a single large network of kinetoplast DNA and that encodes guide RNAs.

Minisatellite DNA. See "Satellite DNA."

Mitochondrial gene. A gene occurring in the genome of a mitochondrion.

Mitochondrial ribonucleic acid (mtRNA). RNA occurring in mitochondria.

Mitochondrion. The organelle of eukaryotic cells in which occurs the energy-yielding process of aerobic respiration.

Moa. Any of a number of recently extinct, large flightless birds found in New Zealand. Some attained height of 10 or more feet.

Mobile DNA. DNA that is able to move around the genome.

Molecular evolution. Descent with modification of the biochemical characteristics of organisms.

Molecular evolutionary clock. The evidently relatively regular, "clock-like" change of some types of evolving biomolecules over geologic periods of time.

Molecular genetics. The science that deals with the molecular aspects of heredity and variation.

Mollusc. Phylum of animals characterized, in part, by a mantle that secretes spicules or shells and a large muscular foot. Includes, among others, snails, clams, and squid.

Monomer. A chemical compound, usually relatively small, that can be linked to other similar compounds into a larger, multicomponent polymer.

Monophyletic group. A group that includes all descendants of a given ancestor. For example, birds are monophyletic, but reptiles are not, since they do not include either mammals or birds, both of which were derived from reptiles (see "Paraphyletic group").

Motif. A short section of amino acids that has a particular sequence indicative of some particular function in a protein.

mRNA (messenger ribonucleic acid). RNA that carries genetic information from the DNA of chromosomes to protein-synthesizing ribosomes.

mtDNA (mitochondrial deoxyribonucleic acid). DNA occurring in mitochondria.

Multicellular. With reference to eukaryotic plants, animals, and fungi that are composed of many cells.

Multigene family. A group of homologous genes generated by gene duplication events.

Mutation. Any change in the sequence of a region of DNA. The word substitution is reserved for those mutations that become fixed in a population. Some (incorrectly) use mutation and substitution interchangeably.

Mutator gene. Genes that promote mutations, such as replacement of one DNA base pair (for example, an adenine-thymine pair) by another (for example, by a guanine-cytosine pair).

Myoglobin. A protein similar to hemoglobin that is abundant in muscle cells and is involved in oxygen storage.

NADP (nicotinamide-adenine dinucleotide phosphate). A coenzyme that plays an important role in intermediary metabolism.

Nanometer (nm). A unit of length, one billionth (10^{-9}) of a meter.

Natural selection. Preferential survival of individuals having advantageous variations relative to other members of the population/species. For natural selection to operate, competition for resources (a struggle for survival) and suitable variation among individuals needs to exist.

Necessities of life. A carbon source (that is, a source of the biogenic elements carbon, hydrogen, oxygen, nitrogen, phosphorus, and sulfur) and a source of energy.

Nematode. Any of a particular taxonomic group (the phylum Nematoda) of elongated cylindrical worms.

Neural. With reference to the nervous system of animals.

Neural-Cell Adhesion Molecule (N-CAM). One of several molecules derived from the same gene that holds together the cells in multicellular mammalian tissue.

Neutral allele. An allele that has no advantage or disadvantage to the organism. Thought to be the predominant type of allele.

Neutral gene. A gene that has no advantage or disadvantage to the organism.

Neutral theory of molecular evolution. The theory that most mutational changes in DNA are neither advantageous nor deleterious to the organism.

Nicotinamide-adenine dinucleotide (NAD). A nonprotein coenzyme that occurs in most cells.

Nif **complex.** The enzyme complex, or the genes (*nif* genes) coding for the proteins in that enzyme complex, involved in biological fixation of molecular nitrogen (N_2).

nif **genes.** See "*Nif* complex."

Nitrate. NO_3^-, a biologically usable form of nitrogen.

Nitrogen fixation. The biological process carried out by various eubacterial and archaebacterial prokaryotes by which molecular nitrogen (N_2) is combined with hydrogen to produce ammonia, a biologically usable form of nitrogen.

Nitrogenase enzyme complex. The enzyme complex involved in biological fixation of molecular nitrogen.

Nitrogenous base. Any of the nitrogen-containing purines (adenine and guanine) and pyrimidines (thymine, cytosine, and uracil) that occur in DNA and RNA.

Noncoding DNA. A portion of a DNA molecule that does not code for a protein or RNA product.

Non-oxygen-producing photosynthesis. Anoxygenic photoautotrophy, such as that carried out by photosynthetic bacteria.

Nonsexual. With reference to organisms that lack capability to reproduce sexually (sexual reproduction involves meiosis and syngamy).

Notochord. A characteristic of chordates: A stiff phosphatic rod that functions to preserve body shape during locomotion in some species. In most vertebrates the backbone performs this function.

Nucleic acid. The genetic information-containing organic acids DNA and RNA.

Nucleotide. Any of several compounds that consist of a sugar (ribose or deoxyribose) linked to a purine (adenine [A] or guanine [G]) or a pyrimidine (thymine [T], cytosine [C], or uracil [U]) nitrogenous base and to a phosphate group and that are the basic structural units of RNA and DNA.

Nucleus. In eukaryotes, a membrane-enclosed organelle that contains the chromosomes.

Occipital bone. A compound bone that forms the posterior, basal part of the skull.

Oligonucleotide. A short, single-stranded DNA molecule; such molecules, consisting of a known sequence of nucleotides, can be synthesized in the laboratory by chemical methods.

One gene–one enzyme theory. The theory that each gene in DNA contains the information required for the biosynthesis of one (protein) enzyme.

Open reading frame (ORF). Any stretch of the DNA that codes for amino acids without any termination (stop) codons. Such a sequence is potentially translatable into a protein or RNA.

Organ. A part of a multicellular organism having a definite form and structure that performs one or more specific functions, such as a leaf in a plant or a limb, or the heart in an animal.

Organelle. A specialized membrane-bound structure in a cell, such as a mitochondrion or a plastid, that performs a definite function.

Orthologous. When comparing genes that belong to the same multigene family, orthologous genes are those that are representatives of the same particular copy of the ancestral gene in different species. For example, if a gene family consists of an A and B copy, the A copies in different species are orthologues, as are the B copies.

Oscillatoriaceae. A taxonomic family of simple, nonheterocystous, filamentous cyanobacteria.

Osteocalcin. A protein found in vertebrate bone that is thought to control the site of deposition of hydroxyapatite, the mineral phase that provides bone with its hardness.

Outgroup. A species, or group of species, used to root an evolutionary tree.

Oxidation. A chemical reaction in which oxygen combines with another substance (or in which hydrogen or electrons are removed from a reacting compound). Energy in the form of heat is released.

Oxygenic. With reference to the presence or production of oxygen.

Ozone. A triatomic form of oxygen, O_3, formed naturally in the upper atmosphere.

Paleozoic Era. The earliest of the three Eras that make up the Phanerozoic Eon of Earth's history, extending from approximately 545 million years ago to approximately 245 million years ago.

Pan-editing. A process in which RNAs are extensively edited, mediated by several overlapping guide RNAs.

Paralogous. When comparing genes that belong to the same multigene family, paralogous genes are those that are representatives of different copies of the ancestral gene in different species. For example, if a gene family consists of an A and B copy, the A copy in one species is paralogous to the B copy in another species.

Paraphyletic group. A group that includes some but not all descendants of a given ancestor. For example, reptiles are paraphyletic, since they do not include either mammals or birds, both of which were derived from reptiles (see "Monophyletic group").

PCR. See "Polymerase Chain Reaction."

PCR primers. Short pieces of single-stranded DNA (oligonucleotides) designed to pair with known regions of the genome to form the necessary double-stranded stretches of DNA for PCR amplification of the region between two primers.

Petrify. The process by which organic-walled organisms are preserved as fossils, embedded three-dimensionally, commonly in a siliceous or calcitic mineral matrix.

Petrographic thin section. A piece of rock sliced and ground sufficiently thin that light can be transmitted through it.

Phage. See "Bacteriophage."

Phanerozoic Eon. The latest of the three principal divisions (eons) of geologic time, extending from the beginning of the Cambrian Period of Earth's history (approximately 545 million years ago) to the present.

Phenotype. The visible characteristics of an organism, in contrast with the genetic makeup (the genotype). The result of the interaction of its genetic constitution and the environment.

Phenylalanine. One of the 20 amino acids commonly occurring in proteins of living systems.

Phosphoglycerate (3PGA). 3-Phosphoglyceric acid, an important intermediate compound in the degradative and biosynthetic pathways of glucose.

Photic zone. The surface layer of oceans or lakes in which sufficient light penetrates to support photosynthesis.

Photoassimilation. Use of light energy to aid the uptake into cells of exogenous organic compounds.

Photoautotrophy. Autotrophy powered by light energy.

Photochemical. Chemical reactions powered by light energy.

Photoheterotrophy. Heterotrophy powered by light energy.

Photomontage. A composite photograph composed of two or more photographs.

Photosynthesis. The biological process carried out by photosynthetic bacteria, cyanobacteria, and plants in which light energy is converted to chemical energy and stored in molecules of synthesized carbohydrates.

Photosynthetic bacteria. Eubacteria capable of anoxygenic photosynthesis.

Photosynthetic space. A surface area or volume within the photic zone in which photosynthesis occurs.

Photosystem I. The later occurring light-requiring step of a two-part sequence of the light reactions of oxygenic photosynthesis.

Photosystem II. The earlier occurring (water-splitting) light-requiring step of a two-part sequence of light reactions of oxygenic photosynthesis.

Phylogeny. Diagram depicting the evolutionary relationships of a group of organisms.

Phylum. A taxonomic category used for major groups of organisms. Organisms assigned to the same phylum tend to have the same biological organization, or body plan. Examples include echinoderms and arthropods.

Phytyl. A 20-carbon isoprenoid hydrocarbon composed of four isoprene monomeric subunits.

Plankton. Organisms inhabiting the surface layer of the sea or lake, consisting of small drifting protists and animals, etc.

Plasmid. An autonomously self-replicating circular piece of DNA, commonly found in bacteria in addition to its own DNA.

Plastid. A general term for any self-replicating organelle in plants and their relatives, including chloroplasts and chromoplasts, for example.

Pleurodire. One of two major lineages of living turtle characterized, in part, by a neck retraction mechanism that draws the neck sideways into the shell.

Point mutation. A potentially heritable chemical change in a gene consisting of a change of one nitrogenous base-containing nucleotide for another.

Poly (A) tail. The 3′ end of an mRNA molecule consisting of a string of adenine residues that has been added after transcription.

Polymer. A multicomponent chemical compound, usually relatively large, that consists of many monomeric subunits that are linked together.

Polymerase. An enzyme that builds polymers from monomeric building blocks.

Polymerase Chain Reaction (PCR). A revolutionary technique that can amplify specific regions of DNA millionfold to billionfold, enabling the sequencing of tiny quantities of DNA that were previously unanalyzable.

Polysaccharide. Any one of a class of carbohydrates whose molecules contain linked monosaccharide (simple sugar) units. Includes starch and cellulose.

Population. A group of conspecific organisms that occupy a relatively well-defined geographic region and show reproductive continuity from generation to generation.

Population genetics. The science that deals with the genetic composition of biological populations.

Posteroanterior. With reference to rear-to-front spatial organization.

Precambrian. A major division of geologic time, extending from the formation of the Earth (approximately 4,500 million years ago) to the beginning of the Cambrian Period of Earth's history (approximately 545 million years ago).

Primary molecular structure. Nucleotide (DNA and RNA) or amino acid (proteins) sequence of macromolecules.

Principle of Frustration. For most organisms, most of the time, there are likely to be many selective forces operating simultaneously. Inevitably, there will be conflicting demands from each of the organism's requirements. The Principle of Frustration operates to find the best compromise solution to these multiple demands.

Prokaryote. Any of diverse types of non-nucleated eubacterial or archaebacterial microorganisms.

Proline. One of the 20 amino acids commonly occurring in proteins of living systems.

Promoter region. Any of various regions of DNA involved in the chemical binding of molecules that are responsible for initiating transcription, that is, gene expression. These are usually considered to be part of the gene they are associated with.

Protein. A polymer composed of a string of amino acid monomers.

Protist. A general term for unicellular eukaryotes, whether plant- or animal-like.

Protozoal. With reference to eukaryotic heterotrophic single-celled organisms.

Pseudogene. A defective copy of a gene, that, because of its defect(s), is unable to be expressed.

Purine. Nitrogenous base found in nucleic acids that contains fused pyrimidine (six-sided ring) and imidazole (five-sided ring) rings. Adenine and guanine are purines.

Purple bacteria. A taxonomic group of Eubacteria that includes, for example, photoautotrophs of the genus *Chromatium*.

Pyrimidine. Nitrogenous base found in nucleic acids that contains a specific arrangement of four carbon and two nitrogen atoms in a six-sided ring. Cytosine, thymine, and uracil are pyrimidines.

Pyrrole. A type of organic compound, a nitrogen-containing five-membered ring that is a principal component of (bacterio)chlorophylls.

Pyruvate. A three-carbon organic compound produced by glycolysis of a six-carbon sugar.

Quaternary molecular structure. The three-dimensional structure of proteins that are made of more than one separate macromolecule. Particularly, the way the subunit chains fit together.

*rbc*L. Abbreviation for the gene that codes for the large subunit that occurs eight times in RUBISCO, a key enzyme for photosynthesis.

*rbc*S. Abbreviation for the gene that codes for the small subunit that occurs eight times in RUBISCO, a key enzyme for photosynthesis.

Reading frame. One of six possible ways of reading a strand of DNA as a series of codons.

Reductionism. A procedure or theory that attempts to reduce complex data or phenomena to simple terms.

Redundancy (of the genetic code). The occurrence of multiple codons that code for the same amino acid.

Reduviid. A taxonomic name for a family of blood-sucking insects.

Regulatory gene. A gene that produces a protein product that controls the expression of another gene.

Release factor. A molecular complex that interacts with stop codons in RNA and terminates protein synthesis by separating the protein from the ribosome.

Reptile. Any of diverse types of commonly scaly, cold-blooded, vertebrate animals such as snakes and lizards.

Restriction enzyme. An enzyme that cleaves DNA molecules, usually at specific nucleotide sequences.

Retrovirus. A virus whose genome is made of RNA that encodes for a reverse transcriptase, an enzyme that uses an RNA template to replicate a DNA copy.

Revertant change. A mutation in DNA that results in a mutated nucleotide being changed back to its original composition.

Rhodoplast. A type of plastid found in red algae (the rhodophytes) that have a red pigment called phycoerythrin, which masks the green color of chlorophyll *a*.

Riboflavin. A nonprotein, growth-promoting member of the vitamin B complex.

Ribonucleic acid (RNA). A single-stranded nucleic acid made up of nucleotides that contain a backbone made of ribose sugar and phosphate (PO_4) molecules and the nitrogenous bases adenine, uracil, guanine, and cytosine.

Ribosomal ribonucleic acid (rRNA). Any of several RNAs that occur in ribosomes.

Ribosome. The intracellular body in prokaryotes and eukaryotes, made of proteins and RNAs, at which protein synthesis occurs.

RNA. See "Ribonucleic acid."

RNA editing. A post-transcriptional modification of the nucleotide sequence of an RNA molecule.

RNA transcript. An RNA molecule that is complementary to the DNA from which it has been copied (transcribed) by RNA polymerase.

rRNA. See "Ribosomal ribonucleic acid."

RUBISCO. The carbon dioxide-fixing enzyme of photosynthesis, ribulose-1,5-bisphosphate carboxylase-oxygenase.

Sacral vertebra. Backbone part of the pelvic region.

Satellite DNA. Short DNA sequences that are repeated hundreds of times in chromosomes. In minisatellite DNA, the repeats are typically dozens of bases long, in microsatellite DNA the repeats are typically only three to seven bases long.

scabrous gene. A gene occurring in mammalian cells that produces products involved in intercellular communication during the formation of nerve tissue.

Schizodemes. Strains of *Trypanosoma cruzi* that contain similar sequences of kinetoplast minicircle DNA molecules.

Second-base position. In a codon, the second of the three positions in the triplet.

Secondary molecular structure. In proteins and nucleic acids, the structure of the molecule brought about by the formation of hydrogen bonds between the amino acids or nucleotides. In proteins, localized structures such as α helices. In single-stranded DNA and RNA, localized double-stranded structures, such as hairpins seen in the SSU rRNA.

Sequence alignment. Matching of homologous sites in the nucleotide or amino acid sequences from a number of homologous genes. Alignment is prerequisite for phylogenetic reconstruction.

Sequence divergence. Differences between nucleotide sequences of DNA or RNA molecules.

Serine. One of the 20 amino acids commonly occurring in proteins of living systems.

Serology. The science dealing with animal fluids (such as blood serum), commonly employing the use of antibodies which recognize specific chemical structures.

Sessile. Nonmobile.

Sexual reproduction. In eukaryotes, the process of reproduction involving formation of spores (in plants) and gametes (in animals) by the process of meiosis, followed by fusion of gametes (syngamy).

Short interspersed element. Short noncoding nucleotide sequences (typically, 300 base pairs long) in DNA that can be repeated up to hundreds of thousands of times at (typically) 1000 base-pair intervals.

Sickle cell anemia. A chronic inherited disease in which the deformed (sickled) blood cells are incapable of normal function.

Silent mutation. A mutational change that alters a codon without altering the amino acid it codes for.

Sister group. The sister group to a particular lineage is the group that is most closely related to that lineage.

Sloth. A type of slow-moving arboreal mammal that today inhabits tropical forests of Central and South America.

Small subunit ribosomal RNA (SSU rRNA). The SSU rRNA is a major component of the smaller of the two subunits that make up the ribosome. This has been a key molecule in reconstructing the Tree of Life. Also called 16S or 18S rRNA.

Somite. A general term for body segment.

Species. The fundamental category of biological classification, ranking below the genus and in some species composed of subspecies or varieties. There are a variety of definitions; commonly used is the *Biological Species Concept*: "Species are groups of actually or potentially interbreeding natural populations, which are reproductively isolated from other such groups."

Stabilizing selection. Selection against phenotypes that deviate from an optimal value of a character.

Stem group. See "Total group."

Stop codon. A codon that terminates transcription.

Strategies of life. Autotrophy and heterotrophy, strategies that evolved to satisfy the necessities of life.

Stratum. A layer of sedimentary rock (sedimentary rocks are formed by the accumulation and consolidation of minerals and organic particles that have been deposited by water, wind, or ice).

Stromatolite. Accretionary organosedimentary structure, commonly finely layered, megascopic, and calcareous, produced by the activities of mat-building microorganisms, principally filamentous photoautotrophic prokaryotes such as various types of cyanobacteria.

Structural gene. A gene that codes for a protein or RNA other than those that serve to regulate the expression of other genes.

Substitution. A point mutation that has become fixed in a population.

Sulfobacteria. An informal name for eocytes.

Syllogism. A piece of deductive reasoning from the general to the particular.

Symbiosis. The living together, in more or less intimate or close association, of two dissimilar organisms.

Symplesiomorphy. Shared primitive characteristic: A characteristic shared by members of a lineage of interest, but also shared by other related lineages.

Synapomorphy. Shared derived character: A unique evolutionary innovation shared by all members of a lineage, but not shared with members of any other lineages. Used by systematists to distinguish different lines of descent in the Tree of Life.

Syngamy. The fusion of gametes (egg and sperm) in sexual reproduction.

Synonymous substitution. A replacement of one nucleotide in a codon by another that codes for the same amino acid.

Synthetic nucleic acid. DNA or RNA made in the laboratory either by chemical synthesis or by use of a bacteriophage RNA polymerase.

Systematics. The study of the classification and history of organisms.

Target gene. Genes that are activated by products of other genes.

Taxonomy. The theory and practice of classifying organisms.

Tertiary molecular structure. In proteins and nucleic acids, the three-dimensional structure of the molecule brought about by its folding upon itself.

Tertiary Period. The earliest of two Periods of the Cenozoic Era of Earth's history, extending from 65 million years ago to 1.6 million years ago.

Testosterone. A steroid hormone that stimulates production of male secondary sex characteristics.

Tetramer. A four-component molecule.

Thermophile. Any of various organisms, such as diverse prokaryotes, that can survive and grow in relatively high-temperature environments.

Thiamine. A vitamin of the B complex that is essential to normal metabolism and is widespread in plants and animals.

Third-base position. In a codon, the third of the three positions in the triplet.

Thoracic vertebra. Rib-bearing backbone part of the upper back region.

Three prime (3′) end, of DNA. See "DNA."

Threonine. One of the 20 amino acids commonly occurring in proteins of living systems.

Thymine (T). One of the four nitrogenous bases in the nucleotides of DNA .

Total group. The total group is the sum of the crown group and stem group; for a group of organisms defined by a particular set of synapomorphies, it includes all living and fossil representatives of the group. The crown group includes only the descendants of the last common ancestor of all living representatives of the group (including extinct lineages). The stem group includes only those lineages of the group whose origins predate the time of origin of the last common ancestor of all living representatives of the group.

Transcript. RNA which has been transcribed (copied) from a DNA template by an RNA polymerase.

Transcription. The process of copying, by base pairing, the DNA nucleotide sequence of a protein coding gene to an RNA nucleotide sequence, called the mRNA.

Transfer ribonucleic acid (tRNA). RNA that by interacting with mRNA at ribosomes adds a specific amino acid to the end of a protein during its synthesis.

Translatable. With reference to mRNAs that contain information that may be "translated" into proteins at ribosomes.

Tree of Life. A tree-like diagram illustrating evolutionary relationships among organisms.

Triassic Period. The earliest of three Periods of the Mesozoic Era of Earth's history, extending from approximately 245 million years ago to approximately 210 million years ago.

Trilobite. Any of diverse types of extinct Paleozoic (245 to 545 million years ago) arthropod characterized by a three-lobed body organization.

tRNA. See "Transfer ribonucleic acid."

Tryptophan. One of the 20 amino acids commonly occurring in proteins of living systems.

Tryptophan synthetase. An enzyme involved in the synthesis of the amino acid tryptophan.

Tunicates. A diverse group of chordates, some of which possess a notochord in their larva.

Tyrosine. One of the 20 amino acids commonly occurring in proteins of living systems.

***Ultrabithorax* gene.** A gene occurring in the homeotic *Bithorax* complex of the fruit fly, *Drosophila*.

Ultraviolet light. That portion of the electromagnetic spectrum lying outside the visible spectrum at its violet end.

Ungulate. Any of a large group of mammals with hooves.

Unit evolutionary period. A measure of the rate of molecular evolution defined as the length of time for a 1% change to occur in the amino acid sequence of a protein.

Uracil (U). One of the four nitrogenous bases in the nucleotides of RNA.

Uridine. A molecule consisting of uracil, one of the four nitrogenous bases in the nucleotides of RNA, and a sugar.

Valine. One of the 20 amino acids commonly occurring in proteins of living systems.

Variable region of a protein. That portion of a protein in which the amino acid sequence differs in different species of organism.

Vertebrate. A subphylum of the phylum Chordata consisting of all animals that possess a bony or cartilaginous skeleton, and a well-developed brain. Includes fishes, amphibians, reptiles, mammals, and birds.

Vertical descent. Refers to the normal transmission of hereditary material from one generation to the next. Usually contrasted with horizontal (lateral) transfer.

Virus. A self-replicating, infectious, nucleic acid-protein complex that requires an intact host cell for its replication. Its genes may be encoded in DNA or RNA.

Visible light. That portion of the electromagnetic spectrum visible to humans and lying between the infrared and ultraviolet portions of the spectrum.

Weathering. The chemical and physical geological processes that disaggregate a rock into its component mineral grains or crystals.

Xenodiagnosis. A method of detection of *Trypanosoma cruzi* parasites in which uninfected reduviid bugs are permitted to take a blood meal from patients and are later examined to determine whether parasites are present in the guts of the reduviids.

INDEX